OPERATIONAL AMPLIFIER CIRCUITS

Brian Moore
John Donaghy

Heinemann : London

William Heinemann Ltd
10 Upper Grosvenor Street, London W1X 9PA

LONDON   MELBOURNE   JOHANNESBURG   AUCKLAND

First published by Pitman Publishing Pty Ltd 1986
First published in Great Britain by William Heinemann Ltd 1987

© Brian Moore and John Donaghy 1986

British Library Cataloguing in Publication Data
Moore, Brian
   Operational amplifier circuits.
   1. Operational amplifiers   2. Integrated circuits
   I. Title   II. Donaghy, John
   621.381'735   TK7871.58.06

ISBN 0 434 91263 8

Printed in Great Britain by
Thomson Litho Ltd, East Kilbride, Scotland

For Angela and Judy
For TK

CONTENTS

PREFACE		viii
INTRODUCTION		ix

Chapter 1
INVERTING AND NON-INVERTING AMPLIFIERS

1.1	Characteristics of an ideal operational amplifier	1
1.2	Definitions	2
1.3	Circuit symbols for differential operational amplifiers	4
1.4	Basic practical operational amplifier configurations: the inverting amplifier	5
1.5	Review of ideal inverting operational amplifier characteristics	6
1.6	Ideal non-inverting amplifier	7
1.7	Summary of some practical operational amplifier parameters	11
1.8	Differential amplifier circuits using operational amplifiers	11
1.9	Summing inverter	12
1.10	The real operational amplifier (introduction)	14
1.11	Additional operational amplifier configurations	14
1.12	Worked examples on multi-input configurations	16
1.13	Practical operational amplifier packages	20
1.14	*Important review points*	21
1.15	Superposition theorem	26
1.16	Nodal analysis	28
1.17	Additional examples	31

Chapter 2
FREQUENCY RESPONSE, SLEW RATE AND BAND WIDTH

2.1	Frequency response of the operational amplifier (introduction)	33
2.2	Operational amplifier: input signal variations (introduction)	35

2.3	Bode approximations (introduction)	36
2.4	Bode approximation: multistage amplifier	38
2.5	Rise time	38
2.6	Small signal bandwidth	39
2.7	Open loop/closed loop relationship for non-inverting amplifier circuits	40
2.8	Slew rate	42
2.9	Slew rate limiting of sine waves	43
2.10	Full power bandwidth (large signal bandwidth)	44
2.11	Slew rate limiting of square wave input	45
2.12	Slew rate limiting of triangular wave input	45
2.13	Worked examples	46
2.14	*Important review points*	47
2.15	Problems for readers	49

Chapter 3
WAVEFORM GENERATORS

3.1	Introduction	51
3.2	Square wave generator	51
3.3	Basic principles of a bistable npn transistor multivibrator	51
3.4	Operational amplifier as a free-running symmetrical multivibrator	53
3.5	Ramp-generator theory	58
3.6	Basic triangular wave generator circuit	59
3.7	Triangular wave generator circuit	59
3.8	Sawtooth wave generator	61
3.9	Introduction to sine wave generators (oscillators)	62
3.10	Oscillators	63
3.11	Wien bridge oscillator	65
3.12	Additional worked examples	65
3.13	*Important review points*	71
3.14	Problems for readers	73

Chapter 4
POWER AMPLIFIERS AND POWER SUPPLIES

4.1	Amplifier classification (introduction)	76
4.2	Ideal Class A direct-coupled stage	76
4.3	Ideal Class A transformer-coupled output stage	78
4.4	Ideal Class B transformer-coupled output stage	79
4.5	Push-pull Class B power amplifier	79
4.6	Crossover distortion	81
4.7	Transistor power relationships for ideal push-pull amplifier circuit	84
4.8	Disadvantages of a transformer as a phase splitter	84
4.9	Complementary symmetry amplifier	84
4.10	Development of a power amplifier using an operational amplifier	85
4.11	Bridge amplifiers	87
4.12	Integrated circuit power amplifiers	90

4.13	Discrete power output stage	91
4.14	Bridge amplifier circuit incorporating IC power amplifiers	92
4.15	Basic heat sink theory	93
4.16	Heat flow model	94
4.17	Introduction to regulators	96
4.18	Basic discrete series regulator circuit	96
4.19	Basic regulator with increased loop gain	97
4.20	Regulators producing output voltages lower than the internal reference element	99
4.21	Pre-regulation	100
4.22	Current limiting circuit	102
4.23	Integrated circuit regulators	102
4.24	Three terminal regulators	103
4.25	Integrated circuit series voltage regulator	103
4.26	*Important review points*	105
4.27	Problems for readers	108

Chapter 5
SELECTED APPLICATIONS OF OPERATIONAL AMPLIFIERS

5.1	Introduction	110
5.2	Audio circuits	111
5.3	Precision rectifiers	111
5.4	High impedance DC voltmeter	112
5.5	Application in medical electronic monitoring systems	113
5.6	Measurement of incident radiation using a photodiode	114
5.7	Full wave ideal rectifier	115
5.8	Peak detector	116
5.9	Variable gain AC amplifier	117
5.10	Differential light intensity circuit	118
5.11	Filter applications	119
5.12	Linear read-out amplifier for resistive bridge circuit	120
5.13	Miscellaneous circuit functions of operational amplifiers	121
5.14	Additional worked examples	124

INDEX   133

PREFACE

This handbook provides a single source of information covering the basic principles of operational amplifier circuits. It will interest both the new and the experienced user of operational amplifiers.

We assume that readers possess a basic knowledge of simple electrical networks and of how transistors operate, but that they have no particular knowledge of operational amplifiers.

In some cases, circuit diagrams and symbol allocations are slightly unconventional. This is a deliberate attempt to simplify the information for the reader. In order to make some circuit diagrams clearer, supply connections have not always been shown.

Acknowledgment is made to Hans Heuer for his constructive critical comments.

Brian Moore
John Donaghy
September 1985

INTRODUCTION

The initial concept of the operational amplifier comes from the field of analogue computers. The term *operational amplifier* now applies to a very high gain, differential input, direct coupled amplifier whose operating characteristics are determined by external feedback elements. By altering the nature of the feedback elements, different analogue operations can be achieved. The important point is that the overall circuit characteristics are, for the majority of practical purposes, determined only by these feedback elements.

Widespread use of integrated circuit operational amplifiers did not really commence until the 1960s. However, within a few years, they have become a standard design tool for applications far beyond the original scope of analogue computer circuits.

Chapter 1
INVERTING AND NON-INVERTING AMPLIFIERS

1.1 Characteristics of an ideal operational amplifier

The best approach to understanding the ideal operational amplifier is to consider the terminal characteristics of the amplifier in 'black box' form. The operational amplifier will be treated in this sense and, in general, what is inside the box will be disregarded.

It is convenient to assume that the amplifier has certain ideal characteristics. The effects of departures from the ideal exhibited by real operational amplifiers may then be expressed in terms of the errors to which they give rise. The ideal amplifier is assumed to have the following characteristics:

a The differential voltage gain is infinite ($A_{V0} = \infty$). This will, in general, make the performance entirely dependent on input and feedback networks.

b The input resistance is infinite ($r_{in} = \infty$). This will ensure that no current flows into the amplifier input terminals and the source network is not loaded.

c The output resistance is zero ($r_o = 0$). Ensures that the amplifier performance is not affected by the load.

d The bandwidth is infinite ($BW = \infty$). This is a bandwidth extending from zero to infinity, ensuring a response to DC signals and no phase change with frequency.

e When the input signal is zero the output signal will also be zero, ie $V_o = 0$ if $V_{in} = 0$.

Since the input impedance is infinite, there will be no current flowing into the amplifier input terminals. In addition, when negative feedback is employed, the differential input voltage reduces to zero. Proof of this is shown in Section 1.14A. These two statements are used as starting points for the analysis of negative feedback type amplifier circuits and will be explored in detail later on.

The equivalent circuit for an ideal operational amplifier (with $r_{in} = \infty$ and $r_o = 0$) is shown in Fig 1.1.

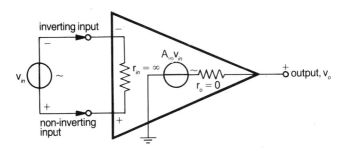

Fig. 1-1 Equivalent circuit for ideal operational amplifier

The amplifier responds only to the difference in voltage between the two input terminals. A positive-going signal at the inverting (−) input produces a negative-going signal component at the output, whereas the same signal at the non-inverting (+) input produces a positive-going output signal component. Essentially, the output signal is the combined response to such input signals. With an input voltage V_{in} the output voltage V_o will be $A_{VO} V_{in}$, where A_{VO} is the differential gain of the amplifier.

1.2 Definitions

The following definitions should be referred to when appropriate.

- **Differential amplifier**

A differential amplifier is an amplifier designed to amplify the difference between two signals. Ideally:
$V_o = (+A_d)V_1 + (-A_d)V_2$ hence $V_o = A_d(V_1 - V_2)$ where A_d = differential gain.

Inverting and non-inverting amplifiers

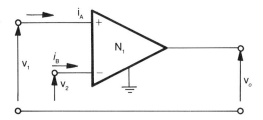

Fig. 1-2 Differential input, single ended output

- **Differential input signal (V_d)**

Referring to Fig 1.2, $|V_d = V_1 - V_2|$ or $|V_d = (V^+) - (V^-)|$ where (V^+) is the signal at the non-inverting input, relative to ground potential, and (V^-) is that at the inverting input.

- **Common-mode input signal (V_c)**

$V_c = \frac{1}{2}(V_1 + V_2)$, the average of the two input signals.

- **Differential gain**

This is the ratio: $\frac{\text{output voltage}}{\text{differential input voltage}}$ when the common-mode input signal is zero.

Hence, $A_d = \frac{V_o}{V_d}$ when $V_c = 0$, ie when $(V_1 = -V_2)$

- **Common-mode gain (A_c)**

$A_c = \frac{V_o}{V}$ when $V_d = 0$, ie when $(V_1 = V_2)$

- **Common-mode rejection ratio (CMRR)**

A real differential amplifier responds not only to the difference between the input signals, but also to their average. The CMRR indicates how much better the differential amplifier is at amplifying differential signals than common-mode signals. That is, the CMRR is a figure of merit for the differential amplifier. The CMRR should be high, so that:
$V_o = A_d V_d + A_c V_c \simeq A_d V_d$ for many practical applications.

3

$$\text{CMRR} = \left|\frac{A_d}{A_c}\right|$$

$$\text{CMRR (dB)} = 20 \log_{10}\left|\frac{A_d}{A_c}\right|$$

$$= A_d(\text{dB}) - A_c(\text{dB})$$

1.3 Circuit symbols for differential operational amplifiers

There are three main varieties of single-ended and differential connections, as shown in Fig 1.3.

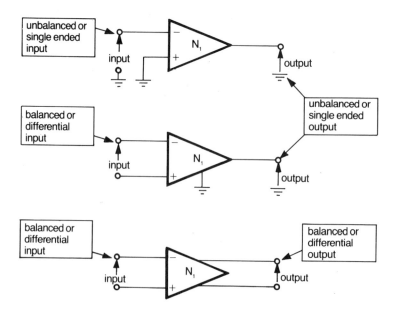

Fig. 1-3 Circuit symbols for operational amplifiers

1.4 Basic practical operational amplifier configurations: the inverting amplifier

Operational amplifiers can be connected in two basic amplifying configurations:
- Inverting
- Non-inverting

The circuit of Fig 1.4 is one of the most widely used operational amplifier circuits. It is an amplifier whose closed-loop gain (gain of amplifier with feedback) from V_{in} to V_o is set by R_1 and R_2. It can amplify AC or DC signals. To understand how this circuit operates under ideal conditions, we use the two assumptions referred to in Section 1.1:

a The voltage V_d between (+) and (−) inputs tends to 0, as $A_d \to \infty$ (from definition 1.2(a)).

b The current drawn by both the (+) and the (−) input terminals is zero.

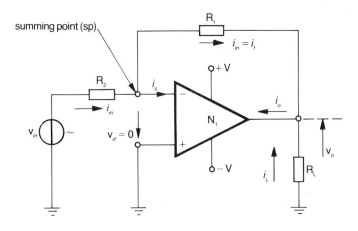

Fig 1-4 Ideal inverting operational amplifier configuration

It is worth pointing out that this circuit only functions properly when the junction of the two resistors R_1 and R_2 goes to the (−) or inverting input of the amplifier. This provides negative feedback, and only negative feedback gives a 'stable amplifier'.

1.5 Review of ideal inverting operational amplifier characteristics

a Since the ideal operational amplifier has infinite gain, it can develop a finite output voltage, V_o, with zero differential input voltage, V_d (Fig 1.4).

b The summing point is a virtual ground at the same potential as the (+) input. ($V_d = 0$, $V^+ = 0$, hence $V^- = 0$).

c The differential input to N_1 is $V_d = 0$. If V_d is zero, then the full input voltage, V_{in}, must appear across R_2, because $V^- = 0$, making the current in R_2:

$$i_{in} = \frac{V_{in}}{R_2}.$$

d Since $i_s = 0$ due to infinite input impedance, the input current i_{in} must also flow in R_1 (*Kirchhoff's Current Law*), thus $i_f = i_{in}$.

e The voltage across R_1 is $i_f R_1$, and since $V^- = 0$ this implies that $V_o = -i_f R_1 = -i_{in} R_1 = -\frac{R_1}{R_2} \cdot V_{in}$

The minus sign implies a 180° phase relationship with the input signal, ie inversion.

This is the characteristic voltage gain transfer function of the ideal inverting amplifier.

f The closed loop voltage gain of the amplifier

$$A_{CL} = \frac{V_o}{V_{in}} = -\frac{R_1}{R_2} \left(\text{Note the gain is strictly } \frac{R_1}{R_2}. \right)$$

g The load current, i_L, flows through R_L.

Thus $i_L = -\frac{V_o}{R_L}$

The operational amplifier output current, i_o, is: $i_o = i_{in} + i_L$

h The input resistance seen by the input signal V_{in} is R_2 since (V^-) is at zero potential.

Example 1
Given that:

$R_2 = 15\ \text{k}$
$R_1 = 150\ \text{k}$
$V_{in} = 1\ \text{V}$
$R_L = 10\ \text{k}$

Refer to Fig 1.5 and determine:
a i_{in}
b A_{CL}
c V_o
d i_o

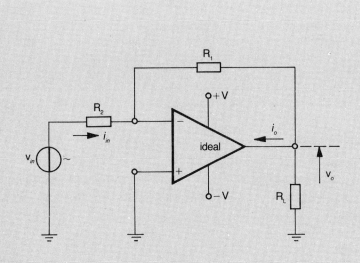

Fig 1·5

Solution

a $\quad i_{in} = \dfrac{V_{in}}{R_2} = \dfrac{1}{15\text{ k}} = 0.07\text{ mA}$

b $\quad A_{CL} = -\dfrac{R_1}{R_2} = -\dfrac{150\text{ k}}{15\text{ k}} = -10$

c $\quad \dfrac{V_o}{V_{in}} = -\dfrac{R_1}{R_2}$, so $\dfrac{V_o}{1\text{ V}} = -\dfrac{150\text{ k}}{15\text{ k}}$ so $V = -10\text{ V}$

d As $\quad i_o = i_{in} + i_L$

where $\quad i_L = \dfrac{-V_o}{R_L} = -\dfrac{(-10)}{10\text{ k}} = 1\text{ mA}$

Hence $i_o = 0.07 + 1 = 1.07\text{ mA}.$

1.6 Ideal non-inverting amplifier

The second basic amplifying configuration of the operational amplifier is the non-inverting amplifier. Fig 1.6 shows a non-inverting amplifier circuit. The output voltage, V_o, has the same phase (or polarity, in the case of DC signals) as the input voltage.

The input resistance of the practical non-inverting amplifier is very large, often over 100 M, and depends on the actual closed loop gain. This very high input resistance, compared to the inverting circuit, is often a reason for choosing this arrangement. An analysis follows:

Operational Amplifier Circuits

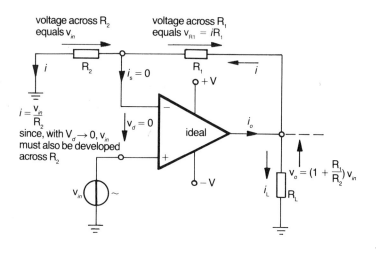

Fig. 1-6 The non-inverting amplifier

$$V^- = V^+ = V_{in}, \text{ so } i = \frac{V_{in}}{R_2}$$

$$V_{R_1} = iR_1 = \frac{R_1}{R_2} \cdot V_{in}$$

$$V_o = V^- + V_{R_1} = V_{in} + V_{R_1}$$

$$\therefore \quad V_o = V_{in} + \frac{R_1}{R_2} \cdot V_{in}$$

so $\quad V_o = \left\{1 + \dfrac{R_1}{R_2}\right\} V_{in}$

and $A_{CL} = \dfrac{V_o}{V_{in}} = 1 + \dfrac{R_1}{R_2}.$

Another method of proving this relationship is shown in Section 1.14B.

Example 2
Determine, for the non-inverting amplifier shown in Fig 1.7,
a the output voltage
b the output current

Solution

a Output voltage $V_o = \left\{1 + \dfrac{R_1}{R_2}\right\} V_{in} = \left\{1 + \dfrac{20\text{ k}}{5\text{ k}}\right\} 2$

    Hence: output voltage $= +10$ V

b $i = \dfrac{V_{in}}{R_L} = \dfrac{2}{5\text{ k}} = 0.4$ mA and $i_L = \dfrac{V_o}{R_L} = \dfrac{10}{5\text{ k}} = 2$ mA

Hence: output current $I_o = 0.4$ mA $+ 2$ mA $= 2.4$ mA

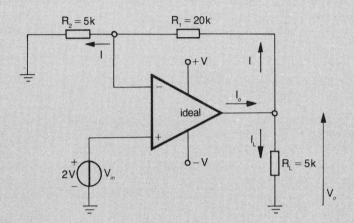

Fig 1·7

Example 3

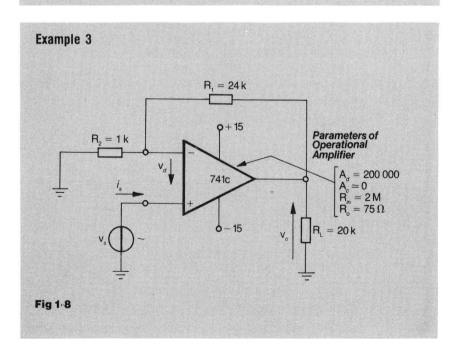

Fig 1·8

Operational Amplifier Circuits

For the circuit shown in Fig 1.8, containing a real operational amplifier whose characteristics are given, determine the approximate value of the input signal current if v_o is 4 volts (rms).

Solution

$$v_d = \frac{v_o}{A_d} = \frac{4}{200\,000} \text{ volts (rms)} = 20\ \mu\text{V (rms)}$$

$$i_s = \frac{v_d}{R_{in}} = \frac{20\ \mu\text{V}}{2\text{M}} = 10\ \text{pA (rms)}$$

Example 4
If v_o is 8 V (pk-pk) at 2 kHz for the 0.2 V input shown in Fig 1.9, determine R_1, R_2, v_d and the voltage across R_2, v_{R_2}, given that the total load placed on the operational amplifier by the feedback resistors is to be 20 k.

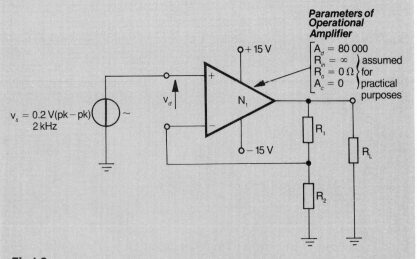

Fig 1-9

Solution

$$\text{Gain required} = \frac{8}{0.2} = +40$$

Hence: $\dfrac{R_1 + R_2}{R_2} = 40$. But $(R_1 + R_2) = 20\,000$

so $\dfrac{20\,000}{R_2} = 40$ or $= R_2 = 500\ \Omega$

Hence: $R_1 = 20\,000 - 500 = 19\,500\ \Omega$

$v_d = \dfrac{v_o}{A_d} = \dfrac{8}{80\,000}$ volts (pk-pk) $= 100\ \mu V$ (pk-pk)

v_{R_2} = inverting input voltage $\simeq$ non-inverting voltage since v_d is negligible at 100 μV by comparison with v_s, which is 200 mV

Hence: $v_{R_2} = 0.2$ V or 200 mV (pk-pk)

1.7 Summary of some practical operational amplifier parameters

a The open loop voltage gain, A_{OL}, is the circuit's differential gain without feedback, ie as if the feedback path was made an open circuit. It follows then that $A_{OL} = A_d$ and $V_o = A_{OL} V_d$.

V_o can never exceed positive or negative saturation voltages. For a ± 15 V supply, the saturation voltages would be about ± 13 volts. Because output voltages, V_o, equals the differential input voltage, V_d, multiplied by open loop gain, A_{OL}, it can be seen that to produce saturation

$V_d = \pm\dfrac{13}{200\,000} = \pm 65\ \mu V$ (approx).

V_d should, however, be neglected only if it is very small compared to V_s, the input signal voltage.

b Some typical parameters of a 741C operational amplifier are as follows:
  i Input resistance = 2 M
  ii Output resistance = 75 ohm
  iii CMRR = 90 dB
  iv Large signal voltage gain = 200 000.

1.8 Differential amplifier circuits using operational amplifiers

Refer to Fig 1.10. It can be shown that:

Differential mode gain, $V_1 \neq V_2$: $\dfrac{V_o}{V_2 - V_1} = \dfrac{R_1}{R_2}$.

Note that the differential gain is defined solely by resistors R_1 and R_2. Proof of this is left for the reader. (The analysis is simplified if R_1 is assumed to be equal to nR_2.)

Operational Amplifier Circuits

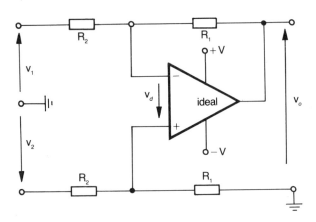

Fig. 1-10 Differential amplifier

1.9 Summing inverter

A summing inverter circuit is used for adding signals, eg as in an audio-mixing circuit. An example of a basic summing inverter circuit is shown in Fig 1.11. This circuit i_{in} is the algebraic sum of the input currents.

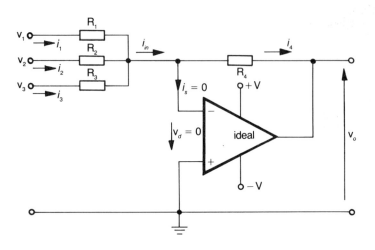

Fig. 1-11 Summing inverter circuit

The (−) input is a virtual earth since (+) input is grounded and v_d is approximately zero.

so $\quad i_1 = \dfrac{v_1}{R_1}, \; i_2 = \dfrac{v_2}{R_2}, \; i_3 = \dfrac{v_3}{R_3}$

Hence: $i_{in} = i_1 + i_2 + i_3 = i_4$, since i_s is relatively small enough to be

12

neglected.
so
$$v_o = V^- - i_4 R_4$$
$$= -i_4 R_4 \text{ (since } V^- = 0)$$
$$= -(i_1 + i_2 + i_3) R_4$$
$$v_o = -\left(v_1 \frac{R_4}{R_1} + v_2 \frac{R_4}{R_2} + v_3 \frac{R_4}{R_3}\right)$$

Example 5
Refer to Fig 1.12.
a Determine the voltage gain applied to each input voltage.
b Calculate the output voltage using Fig 1.12.

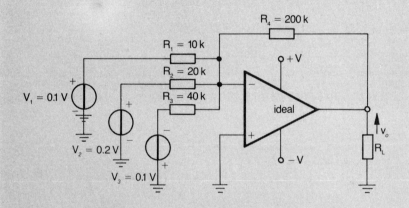

Fig 1·12

Solution

a **i** $A_{CL1} = -\dfrac{R_4}{R_1} = -\dfrac{200 \text{ k}}{10 \text{ k}} = -20$

 ii $A_{CL2} = -\dfrac{R_4}{R_2} = -\dfrac{200 \text{ k}}{20 \text{ k}} = -10$

 iii $A_{CL3} = -\dfrac{R_4}{R_3} = -\dfrac{200 \text{ k}}{40 \text{ k}} = -5$

b $V_o = -\left\{V_1 \dfrac{R_4}{R_1} + V_2 \dfrac{R_4}{R_2} + V_3 \dfrac{R_4}{R_3}\right\}$

$V_o = -[(0·1 \times 20) + (0·2 \times 10) + ((-0·1) \times 5)]$
$= -(2 + 2 - 0·5)$

Hence: $V_o = -3·5$ V

1.10 The real operational amplifier (introduction)

- **Input offset voltage**

Ideal amplifiers produce 0 volts out for 0 volts in. However, in the practical case there may be a small DC voltage at the output even though no differential input voltage is applied.

The 'input offset voltage' may be considered as the DC input voltage which would produce the equivalent DC output voltage. The output voltage is actually caused by differences in the base-emitter junction voltages of the operational amplifier's input differential amplifier.

A related parameter to offset voltage is input offset voltage drift, ie variation in output voltage with temperature. Typical amplifiers usually possess drift levels in the range 5 μV to 40 μV per °C. Low drift amplifiers are available with 10 times better performance.

- **Input bias current and input offset current**

The input bias current is the average of the DC currents required by the amplifier to properly bias its first stage.

I (bias) is defined as: $I_B = \dfrac{I_1 + I_2}{2}$

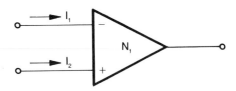

Fig. 1-13 Input bias current

Input offset current is the small difference in bias currents from one input to the other of the operational amplifier. For many applications, input offset current may be ignored. It is only of significance when the output DC level must be accurately related to the input DC level.

- **AC parameters**

Parameter definitions have previously been mainly concerned with magnitudes of DC voltages and currents. Several important AC or frequency dependent parameters will be discussed in chapter 2.

1.11 Additional operational amplifier configurations

- **The integrator**

If a constant current is applied to the input of the integrator it causes the

Inverting and non-inverting amplifiers

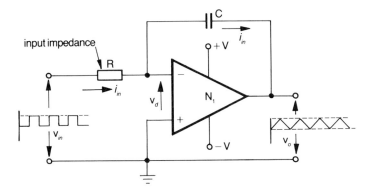

Fig. 1-14 Basic integrator circuit

output voltage of the circuit to decrease linearly with time. The square wave input shown causes the direction of the input current to change sense regularly, resulting in an output voltage which increases, then decreases, linearly with time, ie a triangular waveform.

The output voltage is proportioned to the integral of the input voltage.

$$v_o = -\frac{1}{CR}\int v_{in}\, dt.$$

See Proof in Section 1.14C.

- **The differentiator**

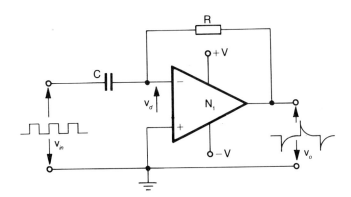

Fig. 1-15 Basic differentiator circuit

The input current is proportional to the rate of change of input voltage with respect to time.

$$v_o = -RC\left(\frac{d}{dt}\cdot v_{in}\right).$$

See Proof in Section 1.14D.

- **The voltage follower**

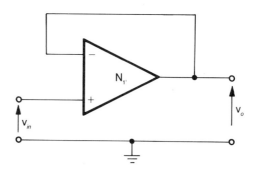

Fig. 1-16 Basic voltage follower circuit

Ideally: Gain = 1
Input Resistance = ∞
$v_o = v_{in}$

See Proof in Section 1.14E.

1.12 Worked examples on multi-input configurations

Example 6

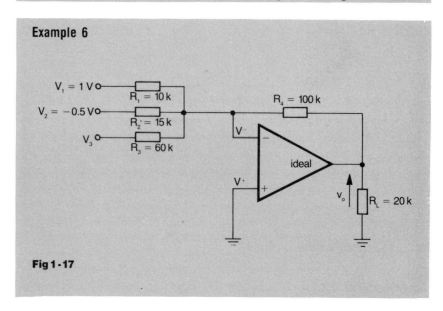

Fig 1-17

Inverting and non-inverting amplifiers

Determine the input, V_3, if the output voltage, V_o, is zero for the circuit diagram shown in Fig 1.17.

Solution

Current in $R_4 = \left(\dfrac{V_1 - V^-}{R_1}\right) + \left(\dfrac{V_2 - V^-}{R_2}\right) + \left(\dfrac{V_3 - V^-}{R_3}\right)$

If V_o is zero, then current in R_4 must also be zero, since $V^- = V^+ = 0$

Hence: $\left(\dfrac{1 - 0}{10\text{ k}}\right) + \left(\dfrac{-0.5 - 0}{15\text{ k}}\right) + \left(\dfrac{V_3 - 0}{60\text{ k}}\right) = 0$

$= \dfrac{1}{10\text{ k}} - \dfrac{0.5}{15\text{ k}} + \dfrac{V_3}{60\text{ k}} = 0$

Multiplying throughout by 60 k: $6 - 2 + V_3 = 0$

Hence: $V_3 = -4$ V

Example 7
Determine I_4, I_0 and I_L for the circuit shown in Fig 1.18.

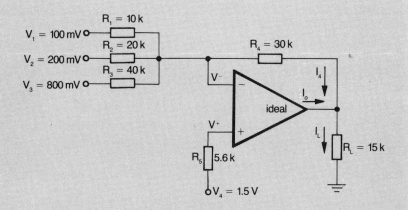

Fig 1·18

Solution

a I_4 = sum of currents in R_1, R_2 and R_3.

$= \left(\dfrac{V_1 - V^-}{R_1}\right) + \left(\dfrac{V_2 - V^-}{R_2}\right) + \left(\dfrac{V_3 - V^-}{R_3}\right)$

But $V^- = V^+ = V_4$ since negligible current will flow in R_5

Operational Amplifier Circuits

so $I_4 = \left(\dfrac{0\cdot 1 - 1\cdot 5}{10\ k}\right) + \left(\dfrac{0\cdot 2 - 1\cdot 5}{20\ k}\right) + \left(\dfrac{0\cdot 8 - 1\cdot 5}{40\ k}\right)$

$ = \left[\left(\dfrac{-1\cdot 4}{10\ k}\right) + \left(\dfrac{-1\cdot 3}{20\ k}\right) + \left(\dfrac{-0\cdot 7}{40\ k}\right)\right]$

$ = (-0\cdot 14 - 0\cdot 065 - 0\cdot 0175)$

Hence: $I_4 = -0\cdot 22$ mA

b $V_o = V^- - I_4 R_4 = 1\cdot 5 - (-0\cdot 2225 \times 30\ k)$
$ = 1\cdot 5 + 6\cdot 675 = 8\cdot 175$ V.

$I_L = \dfrac{V_o}{R_L} = \dfrac{8\cdot 175}{15\ k}$

Hence: $I_L = 0\cdot 545$ mA

c $I_o = I_L - I_4 = 0\cdot 545$ mA $- (-0\cdot 2225$ mA$)$

Hence: $I_o = 0\cdot 77$ mA .

Check

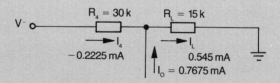

Fig 1·19

From Fig 1.19: $V^- = (I_L R_L) + (I_4 R_4)$
$ = (0\cdot 545$ mA $\times 15$ k$) + (-0\cdot 2225$ mA $\times 30$ k$)$
$ = 8\cdot 175 - 6\cdot 675.$

Hence: $\quad V^- = 1\cdot 5$ volts, which is correct.

Example 8

Calculate V_o, the output voltage of the circuit shown in Fig 1.20.

See Section 1.15 (Superposition theorem).

Solution

Using superposition theorem: $V^+ = \dfrac{V_1 \cdot R_2}{R_1 + R_2} + \dfrac{V_2 \cdot R_1}{R_1 + R_2}$,

ie the sum of the inputs from each source considered separately,

Inverting and non-inverting amplifiers

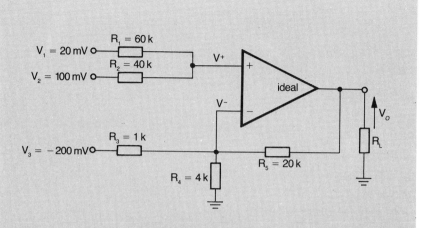

Fig 1-20

$$= \frac{(0\cdot02 \times 40\text{ k})}{100\text{ k}} + \frac{(0\cdot1 \times 60\text{ k})}{100\text{ k}}$$
$$= 0\cdot008 + 0\cdot06$$
$$V^+ = 0\cdot068 \text{ volt}$$

Hence: $V^- = 0\cdot068$ volt

$$\left(\frac{V_3 - V^-}{R_3}\right) + \left(\frac{0 - V^-}{R_4}\right) + \left(\frac{V_o - V^-}{R_5}\right) = 0$$

$$= \left(\frac{-0\cdot2 - 0\cdot068}{1\text{ k}}\right) + \left(\frac{0 - 0\cdot068}{4\text{ k}}\right) + \left(\frac{V_o - 0\cdot068}{20\text{ k}}\right) = 0.$$

Multiplying by 1000 gives:

$$-0\cdot268 - 0\cdot017 + \frac{V_o}{20} - 0\cdot0034 = 0$$

$$\frac{V_o}{20} = 0\cdot0034 + 0\cdot017 + 0\cdot268 = 0\cdot2884.$$

Hence: $V_o = 20 \times 0\cdot2884 = 5\cdot768$ V

Check

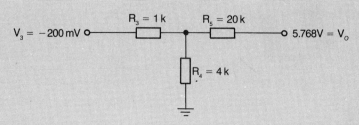

Fig 1-21

From Fig 1.21: $V^- \cdot \left(\dfrac{1}{R_3} + \dfrac{1}{R_4} + \dfrac{1}{R_5}\right) = \left(\dfrac{V_3}{R_3} + \dfrac{0}{R_4} + \dfrac{V_o}{R_5}\right)$

so $V^- \cdot \left(\dfrac{1}{1\,k} + \dfrac{1}{4\,k} + \dfrac{1}{20\,k}\right) = \left(\dfrac{-0\cdot 2}{1\,k} + \dfrac{5\cdot 768}{20\,k}\right)$.

Multiplying by 1000 gives: $V^- \cdot (1 + 0\cdot 25 + 0\cdot 05) = -0\cdot 2 + 0\cdot 2884$

$$V^- = \dfrac{0\cdot 0884}{1\cdot 3}.$$

Hence: $V^- = 0\cdot 068$ V which is correct.

1.13 Practical operational amplifier packages

The LM741 series is a typical general purpose operational amplifier. Details of pin connections for two of its typical packages are shown in Fig 1.22. The exact meaning of some terminal connection applications will be explained in chapter 3. Fig 1.23 indicates the pin connections for a practical inverting amplifier circuit based on the LM741C Operational Amplifier.

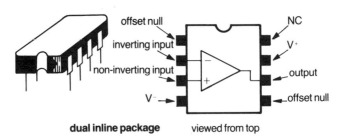

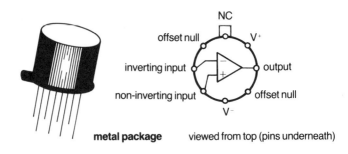

Fig. 1-22 IC operational amplifier packages

Inverting and non-inverting amplifiers

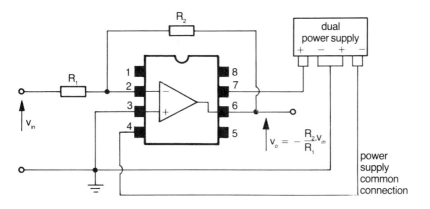

Fig. 1-23 Connections for practical inverting amplifier circuit

The user of operational amplifiers does not require a detailed knowledge of their internal circuitry, but is mainly concerned with the external pin connections.

There is a degree of standardisation in the pin connections with some general purpose operational amplifiers.

1.14 Important review points

A Proof of conditions for v_d to be effectively zero in negative feedback non-inverting amplifier circuits

From Section 1.1.

With an ideal operational amplifier and negative feedback v_d is zero, whereas with a practical operational amplifier and negative feedback, it is normally very close to zero.

Refer to Fig 1.24. Assume the operational amplifier has finite A_{vo} and R_{in}, and zero R_o.

Note that the feedback is negative because the signal at the inverting input substracts from that at non-inverting input, as far as its effect on the output is concerned, and is derived from v_o, which is in phase with v_{in}.

$$v_d = v_{in} - Bv_o \text{ where B is the transfer function of the feedback network.}$$

That is, $v_d = v_{in} - \left(\dfrac{R_2}{R_1 + R_2}\right) v_o.$

But $v_o = A_{Vo} v_d.$

Hence: $v_d = v_{in} - \left(\dfrac{R_2}{R_1 + R_2}\right) A_{Vo} v_d.$

21

Operational Amplifier Circuits

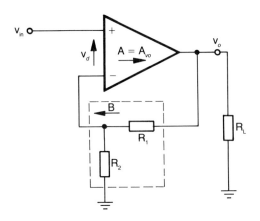

Fig 1-24

$$v_d + \left(\frac{R_2}{R_1 + R_2}\right) A_{Vo} v_d = v_{in}$$

$$v_d \left(1 + \frac{R_2}{R_1 + R_2}\right) A_{Vo} = v_{in}$$

Hence: $v_d = \dfrac{v_{in}}{1 + \left(\dfrac{R_2}{R_1 + R_2}\right) A_{Vo}}$

So $v_d \to 0$ as $\left(\dfrac{R_2}{R_1 + R_2}\right) A_{Vo} \to \infty$.

But $\dfrac{R_1 + R_2}{R_2}$ = closed loop gain. So we need: $\dfrac{\text{open loop gain}}{\text{closed loop gain}} \to \infty$, that is, open gain $\gg$ closed loop gain.

The small size of v_d is of no significance unless it can be shown that it may be neglected by comparison with related signal levels, mainly the input, v_{in} or v_s (Fig 1.25).

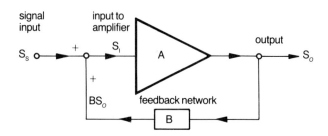

Fig. 1-25 Block diagram of feedback system

To make a more general analysis 'S' is used to represent a signal level which may either be a voltage, or a current level. 'A' is the amplifier gain and 'B' the 'gain' of the feedback network.

$$S_i = S_s + BS_o$$
$$S_o = AS_i$$

Hence:
$$S_i = S_s + ABS_i$$
$$S_i - ABS_i = S_s$$
$$S_i(1 - AB) = S_s$$

$$S_i = \frac{S_s}{(1 - AB)}$$

Hence: $S_i \to 0$ as $-AB \to \infty$

Thus loop gain, AB, needs to have an associated phase shift of 180°, (ie negative feedback) and be very large in comparison with unity, for S_i to approach zero. Under such circumstances, S_i becomes very much smaller than S_s and may be neglected in comparison with it.

B Proof of A_{CL} for non-inverting amplifier (alternate method)

From Section 1.6. Refer to Fig 1.26.

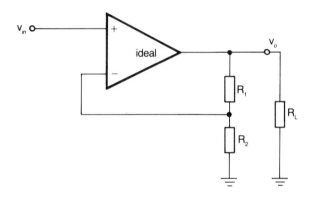

Fig 1-26

$$V^- = V^+ = v_{in}$$

But $V^- = v_o\left(\dfrac{R_2}{R_1 + R_2}\right)$ by potential divider formula (amplifier input current being zero)

so $v_{in} = v_o\left(\dfrac{R_2}{R_1 + R_2}\right)$

Hence:
$$v_o = \left(\frac{R_1 + R_2}{R_2}\right)v_{in}$$

so,
$$v_o = \left(1 + \frac{R_1}{R_2}\right)v_{in}$$

and $A_{CL} = \left(1 + \dfrac{R_1}{R_2}\right),$

where $A_{CL} = \dfrac{V_o}{V_{in}}$

C Proof of the integrator relationship: $v_o = -\dfrac{1}{CR}\int v_{in}\, dt$

From Section 1.11. Refer to Fig 1.27.

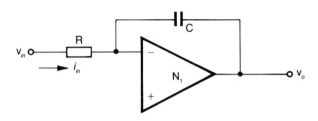

Fig 1-27

The inverting input of the operational amplifier will remain close to ground potential due to the high open loop gain of the amplifier.

The voltage across the capacitor, v_c, is given by:

$$v_c = \dfrac{1}{C} \cdot Q \quad \text{where Q is the instantaneous value of the charge on the capacitor}$$

Now $Q = \displaystyle\int_{t_1}^{t_2} i\, dt$

so $v_c = \dfrac{1}{C}\displaystyle\int_{t_1}^{t_2} i\, dt$

Now $i = \dfrac{V_{in}}{R}$ if inverting input is near ground potential

so $v_c = \dfrac{1}{CR}\displaystyle\int_{t_1}^{t_2} V_{in}\, dt$

now $v_o = 0 - V_c$

so $v_o = -\dfrac{1}{CR}\displaystyle\int_{t_1}^{t_2} V_{in}\, dt$

That is, the output voltage of the circuit is the integral of v_{in} over the time interval t_1 to t_2 seconds, with a constant of proportionality equal to $-\dfrac{1}{CR}$.

- **The differentiator relationship**

D Proof $v_o = -RC\left(\dfrac{d}{dt} \cdot v_{in}\right)$

Refer to Section 1.11. Refer to Fig 1.28.

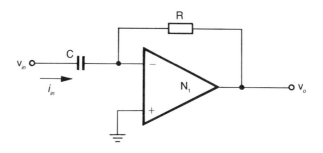

Fig 1-28

$Q = Cv_{in}$, so $\Delta Q = C\Delta v_{in}$,

ie $\quad \int i_{in}\,dt = Cdv_{in}$

or $\quad i_{in} = C\dfrac{dv_{in}}{dt}$,

and since i_{in} also equals current through R and V^- will stay close to ground potential due to the high value of A_{Vo}, it follows that:
$v_o = -i_{in}R$ since V^- is approximately zero.

Hence: $-\dfrac{v_o}{R} = i_{in} = C\left(\dfrac{d}{dt} \cdot v_{in}\right)$

so
$$v_o = -RC\left(\dfrac{d}{dt} \cdot v_{in}\right).$$

That is, v_o is proportional to the time derivative of v_{in} with a proportionality constant equal to $-RC$.

E Proof that the voltage follower circuit has a gain of unity

From Section 1.11. Refer to Fig 1.29.

$V^- = V^+ = v_{in}$
But V^- also $= v_o$, hence $v_o = v_{in}$.

An alternative method would be as follows (from Fig 1.30):

Considering the non-inverting amplifier circuit, the voltage gain
$\dfrac{v_o}{v_{in}} = 1 + \dfrac{R_1}{R_2}$

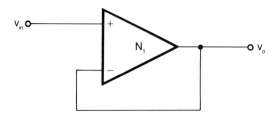

Fig 1-29

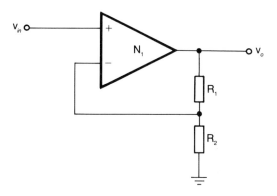

Fig 1-30

In the voltage follower circuit $R_1 = 0$ and $R_2 = \infty$.

Hence: gain $= 1$,

ie the voltage follower circuit is a special case of the non-inverting amplifier circuit.

1.15 Superposition theorem

a In any network containing more than one source (voltage and/or current) the current in, or the potential difference across any branch can be found by considering the effect of each source separately at the appropriate point, and adding the effects.

b Sources of emf not under consideration are replaced by short circuits, while current sources are replaced by open circuits; any internal resistances being left in circuit.

Example 9

Determine the current in R_2 of the circuit shown in Fig 1.31 by using the Superposition Theorem.

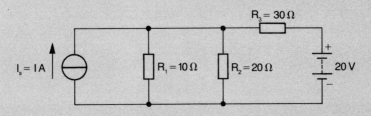

Fig 1-31

Solution

Consider the effect of the current source alone as shown in Fig 1.32.

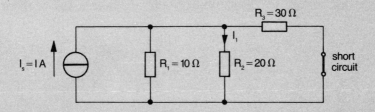

Fig 1-32

R_1 is in parallel with $R_3 = 10\,\Omega \| 30\,\Omega = 7.5\,\Omega$

Hence: $I_1 = \left(\dfrac{7.5}{7.5 + 20}\right) 1\text{A} = 0.273\,\text{A}$

Considering the effect of the voltage source alone, as shown in Fig 1.33:

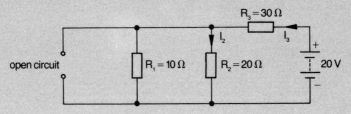

Fig 1-33

R_1 is in parallel with $R_2 = 10\,\Omega \| 20\,\Omega = 6\cdot 67\,\Omega$

Hence: $I_3 = \dfrac{20}{6\cdot 67 + 30} = 0\cdot 545\,A$

Hence: $I_2 = \left(\dfrac{10}{10 + 20}\right) 0\cdot 545 = 0\cdot 182\,A.$

The current in R_2 is the combined value of the currents produced by the current and the voltage source, ie the current in $R_2 = I_1$ (from current source) $+ I_2$ (from voltage source) $= 0\cdot 273\,A + 0\cdot 182\,A = 0\cdot 455\,A$.

1.16 Nodal analysis

a Nodal analysis is a direct application of Kirchhoff's current law and is used to determine the voltage at nodes. Consider the circuit shown in Fig 1.34.

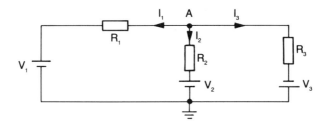

Fig 1·34

b Kirchhoff's current law indicates that $I_1 + I_2 + I_3 = 0$. Now if the voltage at node A is assumed to be V volts, then the currents may be expressed in terms of the voltage drops across the resistors, giving:
$$\dfrac{V - V_1}{R_1} + \dfrac{V - V_2}{R_2} + \dfrac{V + V_3}{R_3} = 0$$

c The currents do not have to be shown all proceeding away from the node. However, the advantage of this arrangement is that V always appears on the left hand side of the equation, making sign changes unnecessary. If the currents are shown going towards the node, V has a negative sign in the equation format.

d If the currents are as shown in Fig 1.35,
then: $I_2 + I_3 = I_1$
from Fig 1.35 it can be seen that:
$$\dfrac{V_2 - V}{R_2} + \dfrac{-V_3 - V}{R_3} = \dfrac{V - V_1}{R_1}$$

Note: This form of equation is disorderly, and so such a choice of current direction is inviting errors to be made.

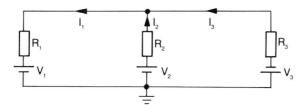

Fig 1·35

Example 10
With reference to the circuit network shown in Fig 1.36, determine:
a The voltage V at the node A relative to ground potential.
b The current in R_2.

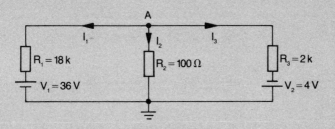

Fig 1·36

Solution

a From Kirchhoff's current law, $I_1 + I_2 + I_3 = 0$

so $\dfrac{V - V_1}{R_1} + \dfrac{V - 0}{R_2} + \dfrac{V - (-V_2)}{R_3} = 0$

Hence: $\dfrac{V - 36}{18\text{ k}} + \dfrac{V}{100} + \dfrac{V - (-4)}{2\text{ k}} = 0$

multiply throughout by 18 k
$V - 36 + 180\,V + 9(V + 4) = 0$
$V - 36 + 180\,V + 9\,V + 36 = 0$, hence $V = 0$

b As the voltage at node A = 0, then the potential difference across $R_2 = 0$, hence the current in $R_2 = 0$.

Example 11
With reference to the circuit shown in Fig 1.37, determine the voltage V at node A relative to ground potential.

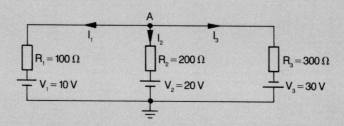

Fig 1-37

Solution
From Kirchhoff's current law, $I_1 + I_2 + I_3 = 0$

thus $\dfrac{V - 10}{100} + \dfrac{V - 20}{200} + \dfrac{V - (-30)}{300} = 0$

multiply throughout by 600

$6V - 60 + 3V - 60 + 2V + 60 = 0$

Hence: $V = 5\cdot 45$ volts

Note: The equations shown in Example 11 may also be expressed in the form:

$$V\left[\frac{1}{R_1} + \frac{1}{R_2} + \frac{1}{R_3}\right] = \frac{V_1}{R_1} + \frac{V_2}{R_2} + \frac{V_3}{R_3}$$

thus $$V = \frac{V_1 G_1 + V_2 G_2 + V_3 G_3}{G_1 + G_2 + G_3}$$

where $G = \dfrac{1}{R}$ = the conductance, measured in siemens.

1.17 Additional examples

Example 12
For the circuit, confirm that the output waveform is as shown in Fig 1.38.

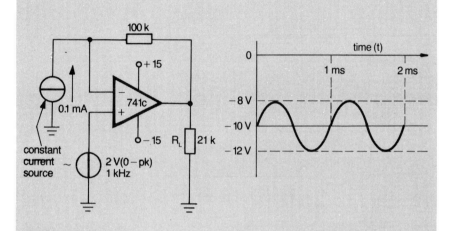

Fig 1·38

Example 13
For the circuit, confirm that the waveforms are as shown in Fig 1.39.

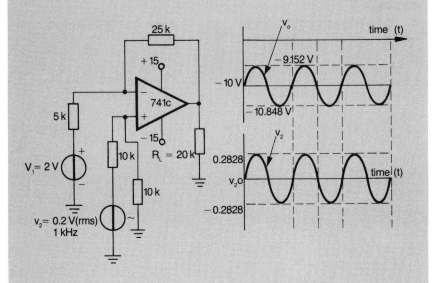

Fig 1-39

Chapter 2
FREQUENCY RESPONSE, SLEW RATE AND BANDWIDTH

2.1 Frequency response of the operational amplifier (introduction)

Many types of operational amplifiers are internally compensated by a small capacitor, eg about 30 pF (for a 741). The internal frequency compensation capacitor prevents the operational amplifier from oscillating with resistive feedback.

Oscillation is prevented by causing the gain of the operational amplifier to decrease at high frequencies. Otherwise there would be sufficient gain and phase shift to cause oscillations. (At a particular high frequency some in-phase signal is fed back to the input, from the output, through the ordinary feedback network.) This particular point is reviewed in more detail in Section 2.14A.

A typical frequency response curve is as shown in Fig 2.1. At low frequencies, the open loop voltage gain is very high, eg a 741 has a gain of about 200 000 times or 106 dB.

Operational Amplifier Circuits

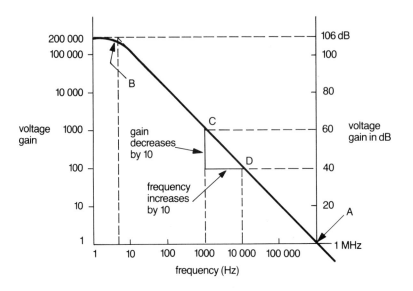

Fig 2·1 Open-loop voltage gain versus frequency

Point B in Fig 2.1 indicates the upper break, or cut-off, frequency. At this point the voltage gain is 0.707 times the gain value at very low frequencies, and corresponds to a halving of power output to a constant load, for constant amplitude input signal.

Points C and D indicate how the gain decreases by a factor of 10 as frequency increases by a factor of 10, for frequencies greater than about ten times, or one decade, above the breakpoint. Also, from Fig 2.1 it can be seen that the voltage gain decreases by 20 dB, ie ten times, for an increase in frequency of ten times (one decade). Hence, the frequency response from B to A is often described as rolling off at a rate approaching 20 dB/decade.

This is equivalent to saying that the gain falls at a rate inversely proportional to the increase in frequency: due to the decreasing reactance of the compensating capacitor referred to earlier.

Point A is the unity gain frequency, and the product of the frequency and gain there is known as the gain-bandwidth product of the amplifier, being (1 MHz × 1) = 1 MHz for the operational amplifier shown.

It can be shown that in the 20 dB/decade roll-off region, the open loop gain at frequency f = $\dfrac{\text{Gain-Bandwidth Product}}{\text{supply frequency } f}$

Example 1

Determine the open-loop gain of an operational amplifier at a frequency of 20 kHz, if it has a unity gain bandwidth of 1·4 MHz.

Solution

From:
$$A_{OL} = \frac{\text{Bandwidth at unity gain}}{\text{frequency}}$$
$$= \frac{1\cdot 4 \text{ MHz}}{20 \text{ kHz}}$$

Hence $A_{OL} = 70$

When an operational amplifier is used as the basis of a negative feedback amplifier, then:

closed loop gain × bandwidth of amplifier =
open loop gain × bandwidth of operational amplifier
ie gain bandwidth product remains constant.

2.2 Operational amplifier: input signal variations (introduction)

The operational amplifier's response depends on the rate of change of the input signals applied to it. It is usual to consider the effect of *sinusoidal signals*, which have no abrupt discontinuity in their waveform, and *square wave signals* which consist of a succession of abrupt changes in slope. It is also usual to distinguish between *low amplitude output signals* (below about 1V peak) and *large signals* (above 1V peak). For small input, and hence output, signals noise is often the limiting factor for undistorted output; for large output signals, slewing rate is frequently the limiting factor.

The gain/frequency response for sinusoidal input signals is represented graphically by a plot of gain in decibels against a logarithmic frequency scale.

Strictly speaking,

$$\text{gain in dB} = 10 \log_{10}\left(\frac{P_o}{P_{in}}\right)$$

where P_o = output power and P_{in} = input power.

From the above definition, $\text{dB gain} = 10 \log_{10}\left(\frac{V_o^2/R_L}{V_{in}^2/R_{in}}\right)$

$$= 20 \log_{10}\left(\frac{V_o}{V_{in}} \cdot \frac{R_{in}}{R_L}\right)$$

Operational Amplifier Circuits

Accepted usage is only strictly correct if $R_{in} = R_L$. It is nevertheless a convenient method of expressing and comparing voltage gains, eg a gain of 1 000 000 is equivalent to 120 dB, whereas one of 500 000 is 114 dB, ie 6 dB less.

Example 2

Determine the voltage gain in dB:

a if $\dfrac{V_o}{V_{in}} = 10$, voltage gain in dB $= 20 \log_{10} 10 = 20$ dB

b if $\dfrac{V_o}{V_{in}} = 100$, voltage gain in dB $= 20 \log_{10} 100 = 40$ dB

c if $\dfrac{V_o}{V_{in}} = \dfrac{1}{\sqrt{2}}$, voltage gain in dB $= 20 \log_{10} \dfrac{1}{\sqrt{2}} = -3$ dB.

Note, since power is proportional to V^2, a decrease in gain of 3 dB represents a halving of the output power, whereas a decrease of 6 dB corresponds to a halving of the voltage gain and the reduction of output power to a quarter of its original value.

2.3 Bode approximations (introduction)

Gain/frequency plots are often approximated by a set of straight lines known as *Bode plots*. The significance of Bode plots can be determined by an analysis of Fig 2.2.

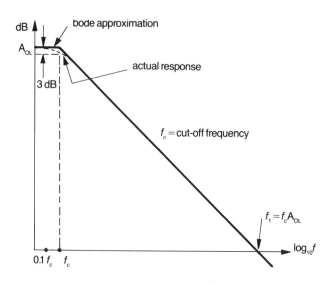

Fig 2·2 Bode approximation (gain/frequency)

Frequency response, slew rate and bandwidth

In Fig 2.2 it is seen that the two straight Bode approximation lines intersect at f_c, and at this frequency the response is 3 dB down. The two straight lines are tangential to the true response at frequencies remote from the cut-off frequency, assuming the absence of other nearby breakpoints.

The corresponding phase/frequency characteristic is as shown in Fig 2.3. It can be seen that the Bode phase plot approximates the phase shift with limits of $0°$ and $-90°$ for frequencies a decade below and above f_c respectively.

A single reactive component will introduce a slope approximating to a magnitude of 20 dB/decade into a frequency response. The slope will be positive or negative, depending upon whether the reactive component is capacitive or inductive and its location in the circuit.

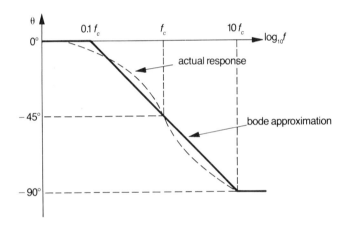

Fig 2-3 Bode approximation (phase/frequency)

It can be shown that: $\theta = -\tan^{-1}\dfrac{f}{f_c}$

for the type of frequency roll-off under discussion.

The Bode frequency plot is useful because, with practice, it enables a good theoretical approximation to the frequency response to be produced quite rapidly.

As far as the response under discussion is concerned:
a when $f \leq 0\cdot1 f_c$, the phase shift due to the breakpoint is effectively zero;
b when $f = f_c$, the phase shift is $-45°$;
c when $f \geq 10 f_c$, the phase shift is approximately $-90°$.
The reader should consult a suitable text for further details.

37

2.4 Bode approximation: multistage amplifier

Bode diagrams are most useful in determining the frequency response characteristics of cascaded gain stages.

The gain of a multistage amplifier is the product of the gains of individual stages. However, since gain is shown logarithmically in Bode plots, the overall Bode response is obtained by adding the responses of the separate stages.

Fig 2.4 illustrates the basic concepts of the production of a Bode plot for cascaded gain stages. The method is applicable to operational amplifier stages having responses similar in their general nature to that of compensated 741 type amplifiers.

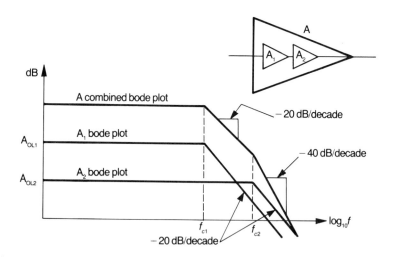

Fig 2-4 Frequency response of cascaded gain stages

Bode diagrams are particularly useful in determining the stability and frequency response of feedback circuits. A frequency dependent response automatically implies that the amplifiers' gain and phase response will both change with frequency. Bode plots are further reviewed in Section 2.14B.

2.5 Rise time

Rise time is defined as the time required for the output voltage to rise from 10% of its final value to 90% of its final value in response to a step input.

This is illustrated in Fig 2.5, and proved in Section 2.14C.

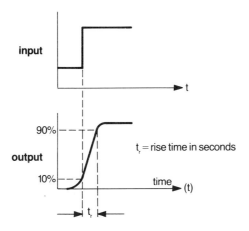

Fig 2-5 Rise time illustration

For a 741 operational amplifier, or any circuit that has a simple -20 dB/decade roll-off above its upper cut-off frequency, it may be shown that

$$\text{upper cut-off frequency} = \frac{0.35}{\text{rise time}}$$

This relationship is very useful as the basis of a practical method of determining the upper cut-off frequency of amplifier circuits. It is thus only necessary to examine the output response to a square wave input to be able to determine the upper cut-off frequency with adequate accuracy.

2.6 Small signal bandwidth

The useful frequency range or bandwidth of an amplifier is defined by an upper frequency limit, f_2, and a lower frequency limit, f_1 (or zero hertz where amplifier is DC coupled).

At f_1 and f_2 the voltage gain is 0·707 its maximum value in the middle of the useful frequency range. This concept is represented in Fig 2.6.

Operational Amplifier Circuits

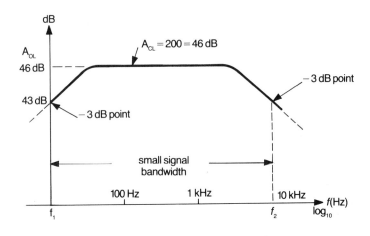

Fig 2-6 Small signal bandwidth

2.7 Open loop/closed loop relationship for non-inverting amplifier circuits

Let f_2 = upper cut-off frequency.
     $f_{2(OL)}$ = upper cut-off frequency of open loop circuit, ie without negative feedback.
     $f_{2(CL)}$ = upper cut-off frequency with negative feedback.
It can be shown that:

closed loop Bandwidth, $f_{2(CL)} = f_{2(OL)} \times \left(\dfrac{A_{OL}}{A_{CL}}\right)$.

Let $R_{o(OL)}$ = output resistance, under open loop conditions, and $R_{o(CL)}$ = output resistance, under closed loop conditions.
It can be shown that:

$$R_{o(CL)} = \dfrac{R_{o(OL)}}{\left(\dfrac{A_{OL}}{A_{CL}}\right)} \text{ and } R_{in(CL)} = R_{in(OL)}\left(\dfrac{A_{OL}}{A_{CL}}\right).$$

The useful relationships above are valid under conditions where the feedback is derived from the output voltage and fed back to the input as a voltage to be added antiphase to the input signal, ie a negative feedback voltage amplifier.

The inverting amplifier obeys the relationship $f_{2(CL)} = f_{2(OL)} \times \left(\dfrac{A_{OL}}{A_{CL}}\right)$ approximately, providing A_{OL} and A_{CL} are gain magnitudes and provided that A_{CL} is > about 10. Strictly speaking $f_{2(CL)} = f_{2(OL)}\left(\dfrac{A_{OL}}{A_{CL}+1}\right)$. R_o is reduced to a low value and $R_{in} = R_i$.

Frequency response, slew rate and bandwidth

Example 3

A non-inverting amplifier circuit has an open loop R_o of 75 Ω and a -3 dB cut-off frequency at 12 Hz. With feedback, R_o decreases to 25 mΩ; determine the amplifier's -3 dB bandwidth.

Solution

$$R_{o(CL)} = \frac{R_{o(OL)}}{\frac{A_{OL}}{A_{CL}}}$$

ie $\quad \dfrac{A_{OL}}{A_{CL}} = \dfrac{R_{o(OL)}}{R_{o(CL)}} = \dfrac{75\ \Omega}{25\ m\Omega} = 3000$

Closed loop bandwidth $= f_{2(OL)} \times \left(\dfrac{A_{OL}}{A_{CL}}\right)$

$\qquad\qquad\qquad\qquad\quad = 12 \times 3000 = 36$ kHz.

Hence: -3 dB bandwidth $= 36$ kHz.

Example 4

An inverting amplifier at constant amplitude input signal delivers a constant output power to its load over the frequency range 10 Hz to 30 kHz. Above 30 kHz the output power gradually decreases until at 350 kHz, for the same input power, the output power is halved, relative to that obtained at 20 kHz. If a 250 mV (rms) input signal at 20 kHz produces a 4 V (pk-pk) output waveform, determine the operational amplifier's upper cut-off frequency, if its open loop gain is approximately 106 dB.

Solution

The implication is that the upper cut-off frequency of the closed loop amplifier must be 350 kHz if the power output is halved.

The closed loop gain of the operational amplifier is given by:

$\dfrac{4\ \text{V (pk-pk)}}{250\ \text{mV (rms)}} = \dfrac{\sqrt{2}\ \text{V (rms)}}{250\ \text{mV (rms)}} = 5{\cdot}656 = A_{CL}.$

Given that $\quad A_{OL} = 106$ dB

$$106\ \text{dB} = 20\ \log_{10}\left|\dfrac{V_o}{V_{in}}\right|$$

$$5{\cdot}3 = \log_{10}|A_{OL}|$$

Hence: $A_{OL} = 199\,526$, say $200\,000$.

Operational Amplifier Circuits

Now $f_{2(CL)} = f_{2(OL)} \times \dfrac{A_{OL}}{(A_{CL} + 1)}$

$\therefore f_{2(OL)} = f_{2(CL)} \times \dfrac{6 \cdot 656}{200\,000}$ (neglecting the '1' since $A_C \gg 1$)

$= 350 \text{ kHz} \times \dfrac{6 \cdot 656}{200\,000}$.

Hence: $f_{2(OL)} = 11 \cdot 6 \text{ Hz}$.

2.8 Slew rate

The slew rate of an operational amplifier indicates how fast its output voltages can change with respect to time. The general purpose operational amplifiers of the 741 series have a maximum slew rate of about $0 \cdot 5$ V/μs, whereas special purpose operational amplifiers may have slew rates of 100 V/μs or more. The lowest slew rate usually occurs at unity gain (unity gain as a result of negative feedback). The limited slew rate capabilities occur because internal compensation circuitry can only change its voltage at a rate limited by associated current sources.

For a capacitor $V = \dfrac{Q}{C} = \dfrac{I \times t}{C}$ for a constant current, I

so $\dfrac{\Delta V}{\Delta t} = \dfrac{I}{C}$

Hence, maximum slew rate $= \dfrac{\text{change in output voltage}}{\text{change in time}} = \dfrac{I}{C}$

Where I = maximum current able to be supplied to the compensating capacitor C.

The essential point is that the capacitor has a maximum charging current associated with it in the operational amplifier, and its maximum voltage change with respect to time occurs when this current is being fully utilised. If capacitor voltage is required to change at a faster rate, there will be distortion of the output signal.

Example 5

A 741 operational amplifier, being used as a voltage follower, has an instantaneous input change of 5 volts. Determine how long it will take for the output voltage to change by 5 volts.

Solution

From slew rate $= \dfrac{\text{change in output voltage}}{\text{time}} = \dfrac{0.5 \text{ V}}{1 \times 10^{-6}\text{s}} = \dfrac{5 \text{ V}}{\text{time}}$

since a voltage follower has unity gain.
So a change of 5 V at output will take 10 μs

2.9 Slew rate limiting of sine waves

Consider the voltage follower in Fig 2.7 where v_{in} is a sine wave of peak amplitude v_p. The maximum rate of change of v_{in}, with respect to time, can be shown to be $2\pi f v_p$ volts/sec where:
a f is the frequency in Hz,
b the peak amplitude is v_p.

The maximum rate of change of voltage with time occurs at the zero crossing points of a sinusoidal waveform; ie the waveform is at its steepest at these points. It should be apparent that increasing amplitude and increasing frequency will cause an increase in steepness at these points. This is because $\dfrac{dv_o}{dt}$ max $= 2\pi f v_p$, that is it is proportional to both f and v_p.

If the required rate of change is larger than the slew rate of the operational amplifier, the output, v_o, will be distorted. If the slew rate is grossly exceeded, then the output waveform will become triangular in nature, as shown in Fig 2.7.

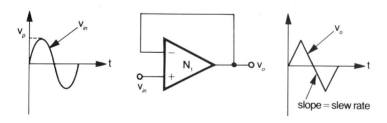

Fig 2-7 Extreme slew rate limiting of sinusoidal output voltage Vo (when $2\pi fVp \gg s$)

The maximum frequency, f_{max}, at which slew rate will not distort the output voltage is:

$f_{max} = \dfrac{\text{slew rate}}{2\pi V_{op}}$ where V_{op} = maximum undistorted output voltage.

Example 6

The typical slew rate of a 741 is 0·5 V/μs. Determine the typical maximum frequency that an output signal may have, at an amplitude of 5 volts peak, without slew rate distortion occurring.

Solution

Since $f_{max} = \dfrac{\text{slew rate}}{2\pi V_{op}}$, ie $f_{max} = \dfrac{0\cdot 5 \times 10^6}{2\pi \times 5}$

then $f_{max} = 16$ kHz .

2.10 Full power bandwidth (large signal bandwidth)

This is the range of frequencies, from zero, over which a specific type of operational amplifier should produce no slew rate distortion of a maximum amplitude sinusoidal output signal.

A 741C amplifier should be capable of producing about 14 V (0-pk) sinusoidal for loads greater than, or equal to, 10 k and has a typical maximum slew rate of 0·5 V/μs. It can be shown that the full power bandwidth (FPB, or large signal bandwidth) is given by: $f = \dfrac{s}{2\pi V_p}$.

Since $2\pi f V_p \leq s$ to avoid slew rate distortion, it follows that $f \leq \dfrac{s}{2\pi V_p}$ is the range of frequencies over which an output signal of amplitude V_p can be obtained without slew rate distortion.

So full power bandwidth for the 741C will be about

$f \leq \dfrac{0\cdot 5 \times 10^6}{2\pi \times 14}$ Hz,

FPB = 5·684 kHz .

Example 7

Calculate the full power bandwidth (FPB) of an operational amplifier operating from ±15 V supply rails, if it saturates within 2 volts of these rails and has a maximum slew rate of 0·8 V/μs.

$\text{FPB} = \dfrac{\text{slew rate}}{2\pi V_{op}} = \dfrac{0\cdot 8 \times 10^6}{6\cdot 28 \times (15 - 2)}$

Hence: FPB = 9·8 kHz

2.11 Slew rate limiting of square wave input

In essence, slew rate limiting is the result of a limit in the ability of the internal circuitry of an operational amplifier to drive capacitive loads, either internal or external. The capacitance that limits the slewing ability is generally the compensation capacitance, although in some instances, it is the load capacitance.

Fig 2.8 shows that at high required rates of signal change the current available to charge or discharge the capacitance is insufficient, and slew rate limiting of the actual output signal occurs.

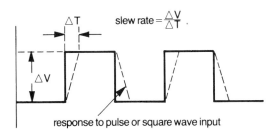

Fig 2-8 The effect of slew rate on a square wave input

2.12 Slew rate limiting of triangular wave input

A triangular wave input will be correctly reproduced, in amplified form, at the output until the repetition rate of the signal is such that the rate of change of voltage inherent in the sloping edge waveform exceeds the slew rate capabilities of the operational amplifier.

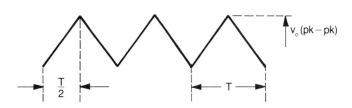

Fig 2-9 Waveform diagram for 2.12

It can be shown that to avoid slew rate distortion:

slew rate $> \dfrac{V_o \text{ (pk-pk)}}{(T/2)}$ is necessary (Fig 2.9), ie slew rate $> 2V_o$ (pk-pk) f_r where f_r = repetition rate.

Hence: $f_r \leq \dfrac{\text{slew rate}}{2V_o \text{ (pk-pk)}}$, to avoid slew rate distortion.

In practice, two effects distort the output waveform. One is the maximum slew rate and the other is the frequency response limitations of the amplifier. The former tends to make triangular type waveforms, whereas frequency response limit tends to introduce an exponential component. At any instant it is the slower of these two effects that determines the actual waveform shape.

2.13 Worked examples

Example 8

A square wave input to an inverting amplifier circuit results in the output waveform shown in Fig 2.10.

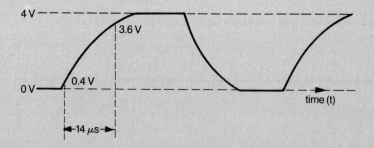

Fig 2-10

The time taken for each positive going output pulse to increase from 0·4 V to 3·6 V is 14 μs. Determine the closed loop upper cut-off frequency (bandwidth).

Solution

The 0·4 V and 3·6 V represent 10% and 90% respectively of the maximum amplitude.
Since the closed loop upper cut-off frequency (see Section 2.5)
$$= \dfrac{0 \cdot 35}{\text{rise time}} \text{ then } f_{CL} = \dfrac{0 \cdot 35}{14 \times 10^{-6}}.$$

Hence: bandwidth or closed loop
upper cut-off frequency = 25 kHz.

Frequency response, slew rate and bandwidth

Example 9

Determine the highest repetition rate of a symmetrical triangular waveform input to an operational amplifier circuit before it suffers from slew-rate distortion.

Date given

Maximum slew rate = 2 V/μs
pk-pk input amplitude = 600 mV
Amplifier gain = 12

Solution

Output waveform is as shown in Fig 2.11.

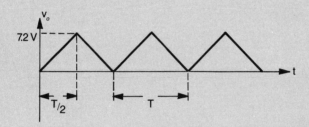

Fig 2-11

v_o(pk − pk) = 12 × 600 mV = 7·2 V
From diagram:

$$\frac{v_{o(pk\text{-}pk)}}{T/2} \leq \text{slew rate to avoid distortion, ie } f_r \leq \frac{\text{slew rate}}{2v_{o(pk\text{-}pk)}}$$

$$= \frac{2}{10^{-6} \times 2 \times 7\cdot 2}.$$

Hence: f_r = 139 kHz.

2.14 Important review points

A Prevention of oscillation

Refer to Section 2.1.

One method of preventing unwanted oscillation is to cause the gain to roll off at high frequencies so that there is insufficient gain available for oscillation to take place at the frequencies where there is the necessary loop phase shift of zero degrees.

Whereas negative feedback reduces the effective input signal to the operational amplifier, positive feedback increases it to a point where the operational amplifier circuit may produce a sinusoidal output signal independent of any external input signal.

B Bode plot

Refer to Section 2.4.

A Bode plot is a graphical method for showing how the gain and phase shift of an amplifier vary with frequency.

A typical plot for an arbitrary RC coupled amplifier is shown in Fig 2.12.

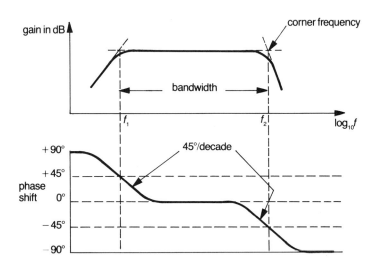

Fig 2·12 Bode plot for an RC coupled amplifier

At low frequencies, due to the coupling capacitors' high reactance, the gain is very low, and the phase angle of the output, relative to the input, is leading by 90°. At frequency f_1 the magnitude of the gain is 3 dB less than its mid-frequency value, and the phase lead is now 45°.

Over the bandwidth the gain and phase shift are reasonably constant. However, at f_2, the gain has again fallen to 3 dB below its mid-frequency value and the phase lag is 45°.

For this amplifier, the phase angle between output and input increases to a maximum of $-90°$. The fall in gain at high frequencies is due to internal active device capacity tending to short the signal to ground, whilst the phase lag is due to the input, output and interstage coupling capacitors.

C Proof of: $f_2 = \dfrac{0\cdot 35}{\text{rise time}}$

Refer to Section 2.5.

Frequency response, slew rate and bandwidth

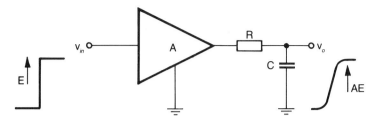

Fig 2-13

An amplifier with a single upper cut-off frequency, f_2, may be represented by the circuit shown in Fig 2.13.

If v_{in} in a step function of amplitude E, the output voltage, v_o, will be:
$v_o = AE(1 - e^{-t/CR})$.

The times t_1 and t_2 taken to reach 10% and 90% respectively of full amplitude will be: $t_1 = CR \ln 0.9$ and $t_2 = CR \ln 0.1$
so 10% to 90% rise time is $t_2 = CR(\ln 0.9 - \ln 0.1)$

Now $f_2 = \dfrac{1}{2\pi CR}$

$= \dfrac{(\ln 0.9 - \ln 0.1)}{2\pi t_2}$

Hence: $f_2 = \dfrac{0.35}{t_2}$.

2.15 Problems for readers

Example 10

An operational amplifier, with open loop cut-off frequency equal to 10 Hz, is used in a non-inverting circuit with closed loop voltage gain of 12. With a square wave input signal, the output rise time is measured as 7 μs. Determine the open loop gain of the operational amplifier, in dB.

Solution

95·6 dB

Example 11

A standard non-inverting amplifier circuit has an output resistance of 30 mΩ. The operational amplifier has an open loop output resistance of 120 Ω, an upper -3 dB cut-off frequency of 10 Hz and an open loop gain of 102 dB. Determine the amplifier's -3 dB bandwidth.

Solution

40 kHz

Chapter 3
WAVEFORM GENERATORS

3.1 Introduction

This chapter deals with operational amplifier circuits which generate signals.

The four common circuits are:
- Square wave
- Triangular wave
- Sawtooth wave
- Sine wave

3.2 Square wave generator

A multivibrator is a circuit which essentially generates square waves.

There are three types of multivibrator:

Astable (free running) — The two states of the circuit are momentarily stable, and the circuit switches repetitively between these states.

Monostable (one shot) — Has only one stable state, but it can be made to change to another state by the application of a suitable trigger pulse. It then returns to the stable state after a time interval determined by circuit component values.

Bistable (flip-flop) — Has two stable states and remains in one until suitably triggered to the other, in which it remains until again triggered to the first state, and so on.

3.3 Basic principles of a bistable npn transistor multivibrator

It is useful to note the basic principles of a bistable npn transistor multivibrator before considering the operational amplifier configurations. First

Operational Amplifier Circuits

consider the transistor as a switch, as shown in Fig 3.1, as the base voltage is changed from reverse bias to forward bias. The resulting voltage waveforms are also shown in Fig 3.1.

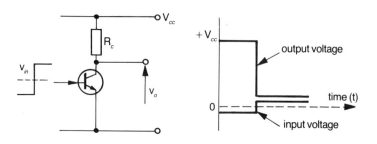

Fig 3-1 Basic principles of a transistor as a switch

Reverse bias Leakage current only in the collector circuit. The voltage drop across R_c is very small. The collector voltage is effectively $_c + V_{cc}$.
Forward bias Collector current flows. The collector voltage falls until it saturates, in this case, at just about the base voltage.

The principle of the above action is now applied to a bistable npn transistor multivibrator circuit as shown in Fig 3.2.

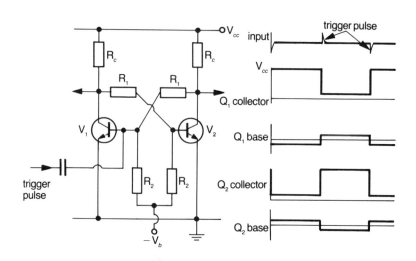

Fig 3-2 Bistable npn transistor multivibrator

Waveform generators

a Initially assume that the base-emitter junction of V_1 is reverse biased. The collector of V_1 will be at V_{cc}. A positive trigger pulse raises V_1 base above zero. V_1 collector voltage falls, causing V_2 base voltage to fall, cutting off V_2. The collector voltage of V_2 rises to V_{cc}.

b A negative trigger pulse lowers collector current of V_1, causing V_1 collector voltage to rise, which raises V_2 base above zero. V_2 collector voltage falls, assisting the fall of V_1 base voltage and resulting in the cut-off of V_1. The collector voltage of V_1 rises to V_{cc}, the initial condition.

3.4 Operational amplifier as a free-running symmetrical multivibrator

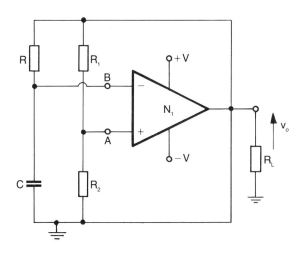

Fig 3-3 Free running symmetrical multivibrator

a Refer to Fig 3.3. The two states of the circuit, between which it switches are:
- positive saturation
- negative saturation

b The output is a square wave of frequency:

$$f = \frac{1}{T} = \frac{1}{2RC \ln\left(\frac{1 + \beta}{1 - \beta}\right)}$$

c The feedback ratio (β) is established by the potential divider made up of R_1 and R_2, ie $\beta = \dfrac{R_2}{R_1 + R_2}$

d The amplitude of the exponential triangular waveform is βV_o

Operational Amplifier Circuits

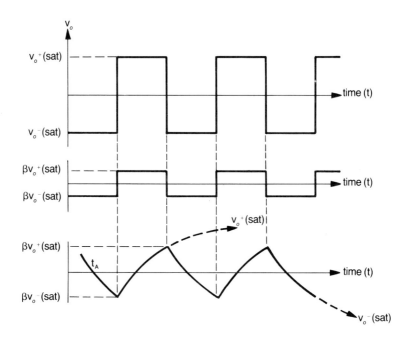

Fig 3-4 Waveforms of free running multivibrator

a   Assume that at time t_A the amplifier output is at V_o^- (sat). The voltage at terminal A (Fig 3.3) is βV_o^- (sat).
b   Simultaneously, at time t_A, terminal B is positive with respect to terminal A and its potential decreases as C discharges through R towards V_o^- (sat).
c   When $V_B = V_A$ (since V_B is changing in potential, whereas V_A is static), the amplifier will start to come out of saturation. The positive feedback to terminal A will then assist in driving the output towards the positive saturation level.
d   The voltage across C cannot change instantaneously. C will now charge up through R towards V_o^+ (sat) and the potential at B will rise exponentially to βV_o^+ (sat).
e   The circuit will start to switch back to the initial state of negative saturation when V_B again equals V_A.

Example 1

Assume that the output of the circuit in Fig 3.5 saturates at some level between +12 V and +14 V.

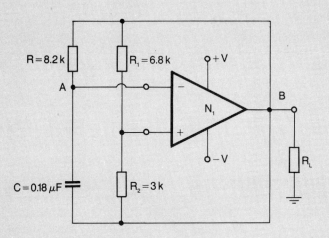

Fig 3·5

a Determine the frequency and amplitude of the waveforms at A and B.
b Determine the possible variation in frequency and amplitude that would be caused by using resistors with tolerance of ±5% and a capacitor with a tolerance of ±10%.

Solution

a
$$\beta = \frac{R_2}{R_1 + R_2} = \frac{3 \text{ k}}{6 \cdot 8 \text{ k} + 3 \text{ k}} = \frac{3 \text{ k}}{9 \cdot 8 \text{ k}} = 0 \cdot 306$$

$$T = 2RC \ln \frac{(1 + \beta)}{(1 - \beta)}$$

Hence: $T = 2 \times 8 \cdot 2 \text{ k} \times 0 \cdot 18 \times 10^{-6} \ln \left(\frac{1 + 0 \cdot 306}{1 - 0 \cdot 306}\right)$

$$= 2 \cdot 952 \times 10^{-3} \ln \left(\frac{1 \cdot 306}{0 \cdot 694}\right)$$

$$= 0 \cdot 002952 \times 0 \cdot 633$$

Hence: $T = 0 \cdot 001867$ seconds and, since $T = \frac{1}{f}$,

$f = 535 \cdot 6$ Hz = frequency at A, and B

The amplitude of the square wave output will be between +12 volts and +14 volts, so the exponential triangular waveform will be β times these values, ie between $12 \times 0 \cdot 306$ and $14 \times 0 \cdot 306$.

Operational Amplifier Circuits

Hence, range is: 3·67 V and 4·29 V (0-pk)

Amplitude of waveform at A between 3·67 and 4·29 V

Amplitude of waveform at B between 12 and 14 V

b $\beta_{(max)} = \dfrac{3\text{k} \times 1·05}{(3\text{k} \times 1·05) + (6·8\text{k} \times 0·95)} = \dfrac{3·15\text{k}}{3·15\text{k} + 6·46\text{k}} = \dfrac{3·15\text{k}}{9·61\text{k}}$

$\beta_{(max)} = 0·328$

$\beta_{(min)} = \dfrac{3\text{k} \times 0·95}{(3\text{k} \times 0·95) + (6·8\text{k} \times 1·05)} = \dfrac{2·85\text{k}}{2·85\text{k} + 7·14\text{k}} = \dfrac{2·85\text{k}}{9·99\text{k}}$

$\beta_{(min)} = 0·285$

so $T_{(max)} = 2 \times 1·05 \times 8·2\text{ k} \times 1·1 \times 0·18 \times 10^{-6} \ln\left(\dfrac{1 + 0·328}{1 - 0·328}\right)$

$= 3·41 \times 10^{-3} \ln\left(\dfrac{1·328}{0·672}\right)$

$= 0·00341 \times 0·681 = 0·00232$ seconds

so $f_{(min)} = \dfrac{1}{0·00232} = 430·9$ Hz

$T_{(min)} = 2 \times 0·95 \times 8·2\text{k} \times 0·9 \times 0·18 \times 10^{-6} \ln\left(\dfrac{1 + 0·2853}{1 - 0·2853}\right)$

$= 2·52 \times 10^{-3} \ln\left(\dfrac{1·2853}{0·7147}\right)$

$= 0·00252 \times 0·587 = 0·00148$ seconds

so $f_{(max)} = \dfrac{1}{0·00148} = 675·1$ Hz .

The frequency is not affected by the square wave amplitude. The square wave amplitude will still be between 12 and 14 volts.

The exponential triangular waveform will be between:
$\beta_{(min)} \times 12$ and $\beta_{(max)} \times 14$
$= 0·285 \times 12$ and $0·328 \times 14$
$= 3·42$ V and $4·59$ V (0-pk) .

Example 2

The basic square wave generator (astable or free running multivibrator) shown in Fig 3.6 is to be constructed so that the 'triangular' waveform is to be half the amplitude of the square wave output. The frequency of the oscillation is to be 200 Hz. Determine the constraints that these requirements place on the component values.

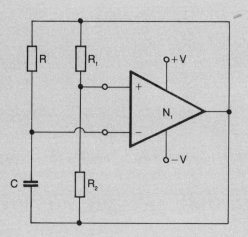

Fig 3-6

Solution

$T = 2RC \ln\left(\dfrac{1+\beta}{1-\beta}\right)$ and the required value of $T = \dfrac{1}{200}$ s = 5 ms

so: $0.005 = 2RC \ln\left(\dfrac{1+\beta}{1-\beta}\right)$ .... (1)

The amplitude relationship requires that $\beta = 0.5$ .... (2)

Substituting (2) in (1) gives: $0.005 = 2RC \ln\left(\dfrac{1.5}{0.5}\right)$

Hence: $0.005 = 2RC \ln 3$
$0.005 = 2.197\, RC$
so $RC = 0.00228$ .... (3)

Also from (2) $\dfrac{R_2}{R_1 + R_2} = 0.5$, so $R_1 = R_2$ is required, and from (3) $R(k\Omega) \times C(\mu F) = 2.28$ is necessary.

A suitable set of values would be:
$R_1 = R_2 = 10$ k
$R = 2.2$ k
$C = \dfrac{2.28}{2.2 \text{ k}} = 1.036\ \mu F$

In practice C would be 1 μF and R either a variable resistor or a variable resistor in series with a fixed resistor. This would enable the frequency to be adjustable to 200 Hz while using standard value components.

Operational Amplifier Circuits

> Some other restrictions on component values would exist to ensure the circuit operated correctly. These would include R and $R_1 \| R_2$ both very much lower than R_{in} of the operational amplifier. R and $R_1 \| R_2$ should be such as to ensure that the output current of the operational amplifier was below its rated maximum output current, or its current limited output, as applicable.

3.5 Ramp-generator theory

It will be useful to note the basic principles of ramp-generator theory, before considering triangular wave generator principles.

Consider a constant current source as shown in Fig 3.7. Capacitor C is charged by the constant current when the switch is closed.

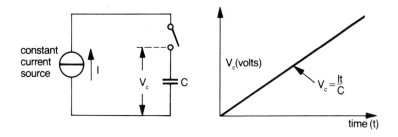

Fig 3-7 Constant-current source ramp generation

V_c is called a *ramp voltage*. If the constant current source is replaced by V_{in} and R_{in} and an operational amplifier, as shown in Fig 3.8, I will equal $\dfrac{V_{in}}{R_{in}}$ since negative input will remain close to ground potential, due to the high open-loop gain of the operational amplifier.

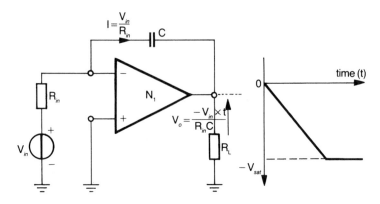

Fig 3-8 The basic integrator circuit as a ramp generator

58

The circuit shown in Fig 3.8 has several disadvantages:
a V_o can only go to $-V$ (sat).
b   When $V_{in} = 0$, $V_o \neq 0$ because the small bias currents will charge the capacitor.
c   To obtain a positive-going ramp, V_{in} must be reversed.

3.6 Basic triangular wave generator circuit

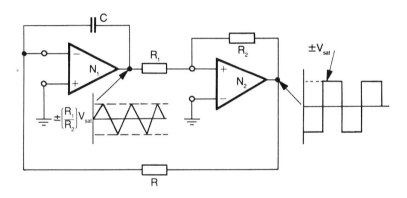

Fig 3-9 Basic triangular wave generator circuit

In Fig 3.9 the size of the triangular amplitude is equivalent to $\dfrac{R_1}{R_2} V_{sat}$ (0-pk).

The frequency of the waveform is given by: $f = \dfrac{R_2}{4R_1 RC}$ Hz

This relationship is proved in Section 3.13A (Important review points).

3.7 Triangular wave generator circuit

Fig 3.10 shows the essential elements of a circuit that combines positive and negative ramp generation to produce a triangular waveform.

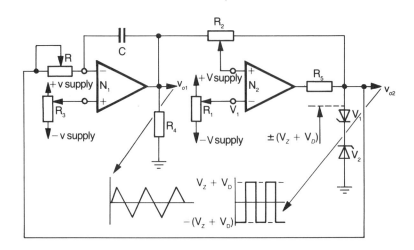

Fig 3·10 Triangular wave generator

a Adjustment of R controls the frequency.
b R_1 applies a voltage, V_1, to the inverting input of N_2. Varying the setting of R_1 alters the DC level of the triangular wave.
c R_2 controls the peak to peak swing of the triangular wave.
d R_3 controls the triangular waveform symmetry, ie the relative rise and fall times of the triangular waveform.
e The zener diode clipping circuit limits the square wave amplitude to $(V_Z + V_D)$ as shown in Fig 3.10: V_D is the zener diode forward voltage drop.
f The feedback capacitor from the output to the inverting input of N_1 means that this circuit can be classified as an integrator. This is consistent with the above description, because the integral of the constant capacitor charging current is a ramp voltage.
(See Section 1.14.)

Example 3

In Fig 3.11 the operational amplifiers saturate at ± 14 V output levels. The reverse biased zener diode voltage drops equal 6·2 V and $V_{D_1} = V_{D_2} = 0·6$ V.
a Determine the frequency of oscillation of the circuit shown.
b State the function of R_1.
c What is the voltage (pk-pk) at point A ?
d What is the voltage (pk-pk) at point B ?
e State the function of the zener diodes V_1 and V_2.

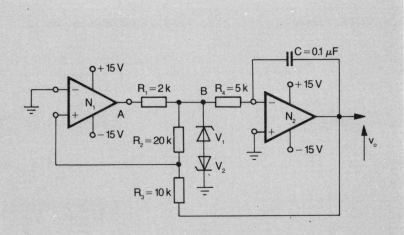

Fig 3-11

Solution

a $f = \dfrac{R_2}{4R_4CR_3} = \dfrac{20\text{ k}}{4 \times 5\text{ k} \times 0 \cdot 1 \times 10^{-6} \times 10\text{ k}} = 1$ kHz

b Essentially, the function of R_1 is to limit the output current of N_1, since when V_A exceeds $(V_Z + V_D)$, the clipping circuit will appear as almost a short-circuit load.

c The peak-to-peak voltage at point A must equal $[+V_{sat} - (-V_{sat})] = 2 V_{sat}$. Hence, the peak-to-peak voltage is $2 \times 14 = 28$ volts.

d The peak-to-peak voltage at point B $= 2(V_D + V_Z)$,

ie B $= 2(0 \cdot 6 + 6 \cdot 2) = 13 \cdot 6$ V

e The zener diodes V_1 and V_2 ensure that N_1 does not saturate heavily, which avoids discrepancies between theoretical and practical frequency values.

3.8 Sawtooth wave generator

The single ramp generator mentioned in Section 3.5 can be modified to make a sawtooth wave generator. A typical circuit is shown in Fig 3.12, where it can be seen that a programmable unijunction transistor (PUT) has been connected across C. The function of the PUT is as follows:

a When the voltage across C = V_p, the PUT short-circuits C by becoming a low resistance from A to K.
b When the voltage across C = V_F, the PUT becomes an open circuit from A to K.

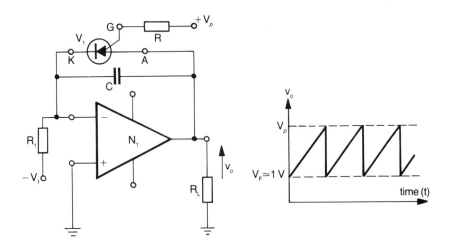

Fig 3-12 Sawtooth wave generator

a The voltage, V_p, applied to the gate terminal of the PUT corresponds to the peak amplitude of the sawtooth waveform.
b The PUT will act as an open circuit from A to K prior to the circuit being energised. C will charge up when the circuit is energised, with the terminal connected to the amplifier output becoming more positive due to the current flowing via R_1 to $-V_1$. This will continue until the potential at A becomes about 0.7 volts higher than that at G, ie 0.7 volts higher than that at V_p. The PUT will then act as a short circuit from A to K.
c C then discharges rapidly until the current from A to K falls below the holding current of the PUT, when the PUT returns to its initial state and the cycle recommences.
d The frequency of oscillation, $f \simeq \dfrac{V_1}{R_1 C V_p}$

3.9 Introduction to sine wave generators (oscillators)

Consider a feedback amplifier in which the feedback network has a gain (B) at one frequency, f_o, as shown in Fig 3.13.

At the same frequency, the amplifier's gain is A and the product, AB, being the loop gain, is exactly unity, which is the highest value of the loop gain at any frequency.

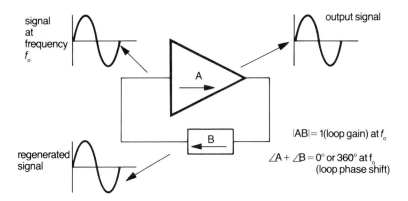

Fig 3·13 Principles of sinusoidal oscillation

a   The amplifier and the feedback network also have the property that, at the same frequency, f_o, as before, the sum of their phase shifts is exactly 0°, or (which amounts to the same thing) 360°. Any signal introduced into the feedback network at the frequency f_o will continue to traverse the feedback loop indefinitely without changing its amplitude, at any specific point in the circuit, even if the initiating signal is removed.

b   Signals at other frequencies circulate, but die out since they would not experience the unity loop gain, or the zero, or 360°, loop phase shift, necessary for their regeneration.

c   The circuit is, in fact, then a signal generator in its own right. If the loop gain is increased to greater than unity at f_o then a signal at that frequency will build up to a usable level when it has traversed the loop a sufficient number of times.

d   By altering appropriate components in the feedback network, the frequency may be easily varied. Feedback networks can be constructed that give the required loss and phase shift combination necessary for oscillation over a wide, tuneable range of frequencies. Hence variable frequency oscillators may be constructed.

3.10 Oscillators

Figure 3.14 shows a typical circuit diagram of a sine wave oscillator using an operational amplifier and a twin T network. In addition, Fig 3.14 also shows the characteristic frequency response of the network. The frequency of oscillation is given by:

$$f_o = \frac{1}{2\pi RC}$$

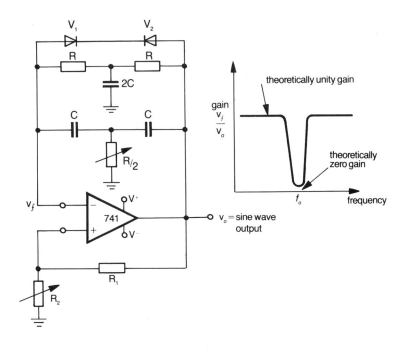

Fig 3·14 Sine wave oscillator and frequency response curve

a A twin T network is used as the feedback network of the oscillator. Since it is the feedback circuit to the operational amplifier's inverting input, as the frequency response shows, there is no negative feedback at the notch frequency and very high negative feedback at all other frequencies.
b Increasing R_2 will increase the amount of positive feedback to the amplifier. It is only at the notch frequency that there is no counterbalancing negative feedback, so that at that frequency the loop gain is relatively high and enables any signals at that frequency to build up to a useful amplitude as they circulate around the circuit.
c V_1 and V_2 tend to stabilise the output signal amplitude since, if it increases in amplitude, their resistance tends to decrease, so increasing the amount of negative feedback and vice versa for decreasing output amplitude.

3.11 Wien bridge oscillator

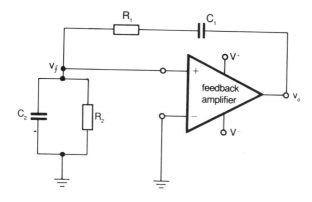

Fig 3-15 Basic circuit of Wien bridge oscillator

It can be shown that if $C_1 = C_2 = C$ and $R_1 = R_2 = R$,

a $v_f = \dfrac{v_o}{3}$ at the oscillation frequency, f_o

b $f_o = \dfrac{1}{2\pi CR}$

These relationships are proved in Section 3.13B.

3.12 Additional worked examples

Example 4

Assume that the operational amplifiers shown in Fig 3.16 saturate at ± 13 V. Describe the effect upon the output signals, v_{o1} and v_{o2}, of adjusting R_{V_1}, from 0 to 10 k.

Operational Amplifier Circuits

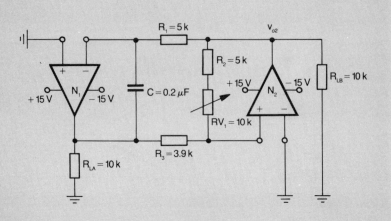

Fig 3·16

Solution

Let $R_{V_1} = 0$. $\quad f = \dfrac{(R_2 + R_{V_1})}{4R_1R_3C} = \dfrac{5000}{4 \times 5000 \times 3900 \times 0\cdot2 \times 10^{-6}}$
$= 320\cdot5$ Hz

Let $R_{V_1} = 10$ k, $f = \dfrac{15\,000}{4 \times 5000 \times 3900 \times 0\cdot2 \times 10^{-6}}$
$= 961\cdot5$ Hz

So as R_{V_1}, varies from 0 to 10 k, the frequency will vary from 320·5 Hz to 961·5 Hz, proportional to $(R_2 + R_{V_1})$.

The amplitude of the triangular waveform, v_{o1}, is:
$$\left(\dfrac{R_3}{R_2 + R_{V_1}}\right) V_{o2}$$
where V_{o2} is a square wave of amplitude ± 13 V (0-pk) at the same frequency.

Hence, amplitude of triangular waveform will vary between
$$\pm\left(\dfrac{3900}{5000}\right) 13 = 10\cdot14 \text{ V}$$
and $\pm\left(\dfrac{3900}{15\,000}\right) 13 = 3\cdot38$ V.

These waveforms are shown in Fig 3.17.

Waveform generators

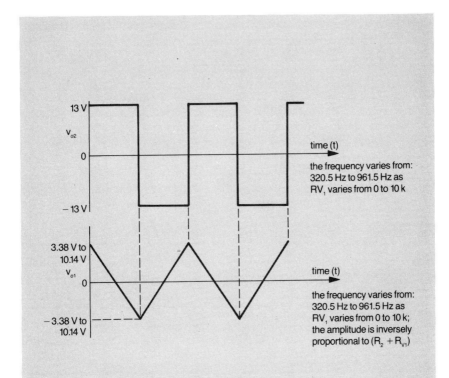

Fig 3-17

Example 5

Assuming that the operational amplifiers in Fig 3.18 go into saturation when their respective outputs get to within 2 volts of their supply rails, determine suitable component values to produce a triangular output signal of amplitude 7 volts (pk-pk) at a frequency of 1·5 kHz. Assume $R_3 = 6·8$ k, and that R_1 and R_2 together are to have a combined value of 10 k.

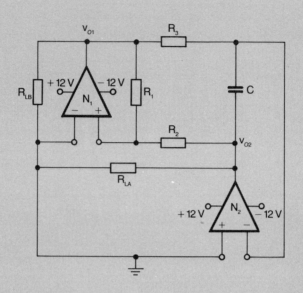

Fig 3-18

Solution

v_{o1} is a square wave output and v_{o2} is a triangular wave output. The frequency of both is given by: $f = \dfrac{R_1}{4R_2R_3C}$

so $\dfrac{R_1}{4R_2R_3C} = 1500$ .... (1)

The amplitude of the triangular wave output is: $\left(\dfrac{R_2}{R_1}\right)V_{o1}$ where v_{o1} will be a square wave of amplitude, $V_{o1} = \pm 10$ v. Considering the given data, a requirement is: $\pm\left(\dfrac{R_2}{R_1}\right)10 = \pm 3\cdot 5$ .... (2)

From equation (1) $\dfrac{R_2}{R_1} = \dfrac{1}{6000\ R_3C}$, and substituting this in equation (2) gives: $\dfrac{10}{6000\ R_3C} = 3\cdot 5$ so, given that $R_3 = 6\cdot 8$ k, $C = 007\ \mu F$

from equation (2) $10\ R_2 = 3\cdot 5\ R_1$ so $2\cdot 857\ R_2 = R_1$.
Since $R_1 + R_2 = 10$ k is given, this means $2\cdot 857\ R_2 + R_2 = 10$ k.

$R_1 = 7\cdot 41$ k
$R_2 = 2\cdot 59$ k

Waveform generators

Example 6

In Fig 3.19 C_1 and C_2 are ganged, matched variable capacitors, with a range covering at least 45 pF to 500 pF each. Determine:

a The nominal loop gain of the circuit at oscillation frequencies.
b The nominal range of frequencies obtainable from the oscillator.
c By considering the equation in Section 3.13B, determine whether the circuit can be guaranteed to oscillate under worst conditions, if the resistors are ±5% tolerance.
d Comment on some of the characteristics that a practical operational amplifier would require, to enable the circuit to function correctly.

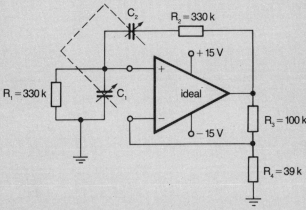

Fig 3·19

Solution

a Amplifier gain is nominally: $1 + \dfrac{R_3}{R_4} = 1 + \dfrac{100\ k}{39\ k} = 3{\cdot}564$ and the loss through the CR network is nominally ⅓. So loop gain is nominally

$$\frac{1}{3} \times 3{\cdot}564 = 1{\cdot}19$$

b Maximum nominal frequency of oscillation is:

$$\frac{1}{2\pi RC_{(min)}} = \frac{1}{2\pi \times 45 \times 10^{-12} \times 330 \times 10^3} = 10{\cdot}72\ \text{kHz}$$

Frequency is inversely proportional to C, and so with C = 500 pF, will be: $\dfrac{45\ \text{pF}}{500\ \text{pF}} \times 10{\cdot}72\ \text{kHz} = 965\ \text{Hz}.$

Hence, the range of frequencies = 965 Hz to 10·72 kHz.

c In cases where $R_1 \neq R_2$ and $C_1 \neq C_2$,

$$f = \frac{1}{2\pi\sqrt{C_1 C_2 R_1 R_2}}$$

and the 'gain' of the CR network is: $\dfrac{1}{1 + \dfrac{C_2}{C_1} + \dfrac{R_1}{R_2}}$

The 'gain' of the network with 5% tolerance resistors will vary between: $\dfrac{1}{1 + 1 + \dfrac{0.95 \times 330\ k}{1.05 \times 330\ k}}$ and $\dfrac{1}{1 + 1 + \dfrac{1.05 \times 330\ k}{0.95 \times 330\ k}}$

ie the '*gain*' of the network will vary between:

$\dfrac{1}{2 + \dfrac{0.95}{1.05}}$ and $\dfrac{1}{2 + \dfrac{1.05}{0.95}}$

$= \dfrac{1}{2.905}$ and $\dfrac{1}{3.11}$

$= 0.344$ and 0.322.

The gain of the amplifier will be between:

$1 + \dfrac{R_3}{R_4} = 1 + \dfrac{0.95 \times 100\ k}{1.05 \times 39\ k}$ and $1 + \dfrac{1.05 \times 100\ k}{0.95 \times 39\ k}$

$= 1 + \dfrac{95\ k}{40.95\ k}$ and $1 + \dfrac{105\ k}{37.05\ k}$

$= 3.32$ and 3.83.

Hence the loop gain will vary between a minimum of 3.32×0.322 and a maximum of 3.83×0.344, ie between 1.07 and 1.32.

Thus the circuit should always oscillate since, with an ideal operational amplifier, the loop gain always exceeds unity. (Even with a practical amplifier with an open loop gain exceeding about 50, the circuit should still oscillate.) The actual frequency range would vary from:

$$\frac{1}{2\pi \times 500 \times 10^{-12} \times 330 \times 10^3 \times 1.05} \text{ to}$$

$$\frac{1}{2\pi \times 45 \times 10^{-12} \times 330 \times 10^3 \times 1.05}$$

$= 918.6$ Hz to 10.2 kHz,

for R_1 and R_2 at the top of tolerance limit.

and $\dfrac{1}{2\pi \times 500 \times 10^{-12} \times 330 \times 10^3 \times 0.95}$ to

$$\frac{1}{2\pi \times 45 \times 10^{-12} \times 330 \times 10^3 \times 0.95}$$

= 1·02 kHz to 11·3 kHz ,

for R_1 and R_2 at the bottom of tolerance limit.

d To avoid slew rate effects, S would have to be at least the maximum value of $2\pi fV$,
ie $2\pi \times 11·3$ kHz $\times 15 = 1·06$ V/µs.

To avoid problems associated with phase shift in the amplifier, the closed loop upper cut-off frequency should be at least about 10 times higher than the highest oscillation frequency, ie $10 \times 11·3$ kHz $= 113$ kHz. This implies a closed loop gain-bandwidth product of 3×113 kHz, ie 339 kHz. Now, assuming the minimum gain-bandwidth product is around $1/10$ of the typical figure for an operational amplifier, this implies that the nominal gain-bandwidth product requirement is about 3·4 MHz, ie an open loop gain of 200 000 and an open loop upper cut-off frequency of 17 Hz, for example.

3.13 Important review points

A Proof of $f = \dfrac{R_2}{4R_1 RC}$

Refer to Section 3.6.

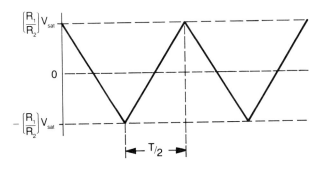

Fig 3·20

Operational Amplifier Circuits

Total voltage change in time T/2 is:

$$\left(\frac{R_1}{R_2}\right)V_{sat} - \left[-\left(\frac{R_1}{R_2}\right)V_{sat}\right] = \left(\frac{2R_1}{R_2}\right)V_{sat} \quad \ldots \ldots (1)$$

$V_C = \dfrac{Q}{C} = \dfrac{It}{C}$ (Where $I = \dfrac{V_{sat}}{R}$, due to negative input of N_1 (Fig 3.9) remaining close to ground potential because of high open loop gain)

so: $V_C = \dfrac{V_{sat} \times t}{RC} \quad \ldots \ldots (2)$

In this case $V_C = \left(\dfrac{2R_1}{R_2}\right)V_{sat}$

ie $\left(\dfrac{2R_1}{R_2}\right)V_{sat} = \dfrac{V_{sat} \times t}{RC}$ (where $t = T/2$)

so $\dfrac{2R_1}{R_2} = \dfrac{T}{2RC}$ hence $T = \dfrac{4R_1 RC}{R_2}$

$$f = \dfrac{R_2}{4R_1 RC} \text{ Hz}$$

B Proof of: $Vf = \dfrac{V_o}{3}$ and $f = \dfrac{1}{2\pi CR}$

From Section 3.11. The basic circuit of the Wien bridge oscillator is as shown in Fig 3.21.

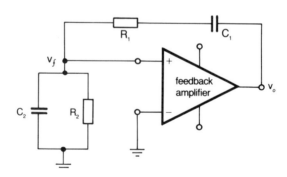

Fig 3-21

The relationship between v_f and v_o is given by the equation:

$$v_f = \dfrac{V_o}{1 + \dfrac{C_2}{C_1} + \dfrac{R_1}{R_2} + j\left(\omega C_2 R_1 - \dfrac{1}{\omega C_1 R_2}\right)}$$

The j term disappears when: $\omega C_2 R_1 = \dfrac{1}{\omega C_1 R_2}$

This occurs when: $\omega = \dfrac{1}{\sqrt{C_1 C_2 R_1 R_2}}$ .... (1)

This leaves $v_f = \dfrac{v_o}{1 + \dfrac{C_2}{C_1} + \dfrac{R_1}{R_2}}$ .... (2)

when $\omega = \dfrac{1}{\sqrt{C_1 C_2 R_1 R_2}}$, v_f is therefore in phase with v_o but with a magnitude determined by the denominator of the relationship shown in equation (2).

If $C_1 = C_2 = C$ and $R_1 = R_2 = R$, then equation (1) reduces to

$$2\pi f = \omega = \dfrac{1}{\sqrt{C^2 R^2}} = \dfrac{1}{CR} \quad \text{ie} \quad f = \dfrac{1}{2\pi CR}$$

Equation (2) reduces to:

$$v_f = \dfrac{v_o}{1 + \dfrac{C}{C} + \dfrac{R}{R}}$$

ie $v_f = \dfrac{v_o}{3}$

Hence, if the amplifier is non-inverting and with a gain of slightly more than 3, the circuit will function as an oscillator at a frequency $f = \dfrac{1}{2\pi CR}$.

Discrete circuit Wien oscillators may be designed to operate satisfactorily up to about 1 MHz. At higher frequencies the value of C and R become comparable with circuit stray capacity and capacitor loss resistance. Consequently, frequency prediction becomes unreliable.

3.14 Problems for readers

Example 7

In Fig 3.22 V_1 and V_2 are 6·2 V ±5% zener diodes when operated at a reverse current of 5 mA. Their forward voltage drop is nominally 0·65 volts at 5 mA forward current. Determine:

a A suitable value for R_4 if the zener diodes are to clip the output waveform of the operational amplifier to a nominal amplitude of 6·85 volts. Assume that the operational amplifier output saturates at ±13 volts.

b The effect on the frequency and amplitude of the triangular and square wave waveforms if:
  i R_1 is doubled in value,
  ii R_3 is doubled in value.

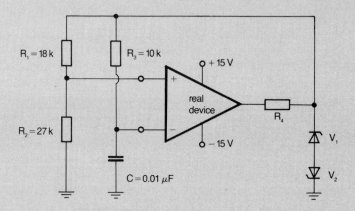

Fig 3-22

Solution

a $R_4 = 1.23$ k.
b The amplitude of the square wave is governed by the zener diode, so it is unaffected. The frequency of both triangular and square waveforms changes from 3·61 kHz to 5·46 kHz. The amplitude of the triangular waveform will decrease from 4·11 V to 2·94 V.
c No effect on the amplitude of the waveforms. The frequencies of both waveforms halve in value,
ie from 3·61 kHz to 1·80 kHz $\left(f \propto \dfrac{1}{R_3} \right)$

Example 8

Assume that the operational amplifier shown in Fig 3.23 saturates when the output reaches the ± 13 V levels.

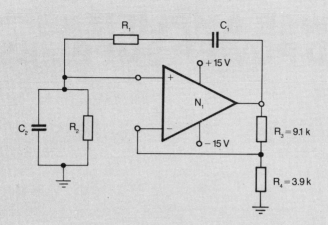

Fig 3-23

a With $R_1 = R_2 = 1.5$ k and $C_1 = C_2 = 0.01$ μF
  i confirm that the circuit is capable of oscillating with nominal value components as shown
  ii check whether the circuit is capable of oscillating in all cases, if resistors are ±5% and the capacitors are ±10%
  iii determine the slew rate requirement for the operational amplifier if slew rate distortion is to be avoided. Assume nominal conditions as in (i).
b If $R_1 = 20$ k and $R_2 = 2$ k with $C_1 = 2200$ pF and $C_2 = 220$ pF
  i determine the necessary nominal gain of the amplifier if the circuit is to oscillate (Assume nominal loop gain is 1.1.)
  ii determine suitable values for R_3 and R_4 if the circuit is to be bias current compensated (For the circuit to be bias current compensated: $R_2 = R_3 \| R_4$.)
  iii determine the nominal frequency of oscillation
  iv what is the nominal slew rate requirement of the operational amplifier now?

Solution
a **i** Loop gain is 1.11 so circuit should oscillate.
  ii Loop gain can vary between 0·935 and 1·314, so in some cases, circuit tolerances may combine to prevent circuit oscillation.
  iii $S \geq 0.87$ V/μs necessary.
b **i** 12·21
  ii $R_3 = 24.4$ k, $R_4 = 2.18$ k
  iii 36·2 kHz
  iv $S \geq 2.96$ V/μs necessary

Chapter 4
POWER AMPLIFIERS AND POWER SUPPLIES

4.1 Amplifier classification (introduction)

Power amplifiers are classified in accordance with the fraction of each cycle of a sinusoidal signal during which the active devices in the output stage are forward biased, ie passing collector current in the case of transistors.

Class A The active device(s) are forward biased throughout the entire period of each cycle of a sinusoidal signal. The signal increases the forward bias during one half of each cycle and decreases it during the other, but never reverse biases the stage.

Class B The active devices are reverse biased for one half of each sinusoidal signal cycle and are forward biased by the signal during the other half. Where the signal's instantaneous amplitude is of importance, as in audio amplifiers, two active devices are required — one to amplify positive going half cycles and the other to amplify negative going half cycles.

Class AB This is a classification for amplifiers operating between the A and B classes, ie where the active devices are forward biased for less than the complete cycle, but for more than half of each cycle.

Class C In Class C operation the collector current will be zero for more than half the input signal cycle. The Class C amplifier has the highest efficiency, of the classes referred to, and is mainly used in radio frequency power amplifiers. It is especially useful where the signal is frequency modulated, but cannot be used to amplify amplitude modulated signals.

4.2 Ideal Class A direct-coupled stage

An 'ideal' amplifier output stage has: no losses due to transistors saturating, no losses due to electrical dissipation within transformers, and no losses due to the addition of extra components to the basic circuit for the purpose of overcoming problems such as transistor non-linearity and temperature dependence, etc.

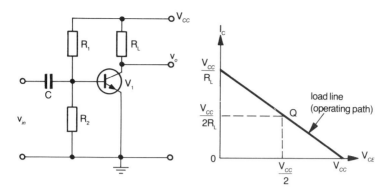

Fig 4-1 Basic ideal Class A direct-coupled circuit and load line

When the circuit in Fig 4.1 is biased in Class A operation:
a The circuit is biased to give equal positive and negative output swings by making V_{CE} at point Q equal to $\dfrac{V_{CC}}{2}$.
b The optimum quiescent collector current value = $\dfrac{V_{CC}}{2R_L}$.
c The transistor, Q_1, must be capable of dissipating $2P_o$ (max) watts, where P_o (max) = $\dfrac{V_{CC}^2}{8R_L}$, is the maximum signal power output to the load R_L.
d The transistor must also be able to withstand V_{CC} volts from collector to emitter and a peak collector current of $\dfrac{V_{CC}}{R_L}$ amperes.
e Output stage efficiency refers to the effectiveness of the output stage of a power amplifier in converting DC input power to the stage into useful signal output power to the load. It is the ratio of useful output power, P_o, to DC input power, P_{CC}, expressed as a percentage. Efficiency
$$\eta = \dfrac{P_o}{P_{CC}} \times 100\%$$
f The maximum efficiency of such a stage is 25% and this occurs at P_o (max).
g The low efficiency is due to the presence of a DC current component in the load. The DC power dissipation is entirely lost, contributing no useful output power component.
h In many cases, the DC current also produces other undesirable effects, especially in the case of electromagnetic loads such as speakers and motors, since the magnetic flux produced by the current may tend to saturate the magnetic circuitry.

4.3 Ideal Class A transformer-coupled output stage

The circuit shown in Fig 4.2 is biased to give equal positive and negative output swings by making the quiescent collector current, I_{CE}, such that $I_{CE} = \dfrac{V_{CC}}{R'_L}$ where R'_L is the reflected load seen by the transistor and equals $n^2 R_L$ where n = transformer primary to secondary turns ratio, n:1.

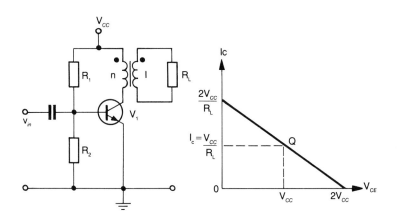

Fig 4-2 Basic ideal Class A transformer-coupled output stage and load line

The ideal circuit action is:
a The reflected collector load impedance, seen by the transistor, is $R'_L = n^2 R_L$.
b The optimum Q point bias current is $I_{CE} = \dfrac{V_{CC}}{R'_L}$
c The transistor must be capable of dissipating $2P_o(\max)$ watts, which occurs at zero input signal conditions, where $P_o(\max) = \dfrac{V_{CC}^2}{2R'_L}$ watts.
d The transistor must also be capable of withstanding $2V_{CC}$ volts from collector to emitter and a peak collector current of $\dfrac{2V_{CC}}{R'_L}$ amperes.
e The maximum efficiency of the stage is 50% and this occurs at $P_o(\max)$. Because the DC component of the collector current is isolated from the load efficiency is improved.
f The only power dissipation in the load is due to the reflected signal component of the collector current.
g The load R_L must not be disconnected from the amplifier when the amplifier is operating, since the increased value of R'_L will cause V_{CE} to greatly exceed $2V_{CC}$ volts.

4.4 Ideal Class B transformer-coupled output stage

The circuit shown in Fig 4.3 when functioning in Class B operation, acts as follows:
a The transistor is biased at cut off, ie the Q point is on the V_{CE} axis.
b If the input signal is sinusoidal in nature, a current flows for only half of each cycle and is zero during each other half cycle.
c Class B is more efficient than Class A because there is no power dissipation in the transistor during one half of each input cycle. When Class B amplifiers are used to amplify amplitude modulated signals, eg in an audio amplifier, two active devices are employed, one to amplify positive-going half cycles and the other to amplify the negative-going half cycles.

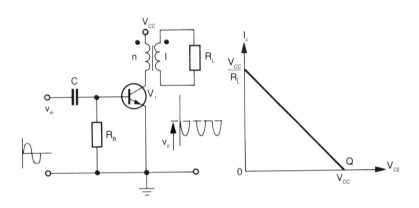

Fig 4-3 Ideal Class B transformer-coupled output stage and load line

A Class AB amplifier operates between Class A and Class B: operating the point lies between cut off and the centre of the load line.

4.5 Push-pull Class B power amplifier

If a Class B amplifier is to reproduce amplitude modulated waveforms correctly, a transformer-coupled, push-pull amplifier may be used. A typical circuit is shown in Fig 4.4.

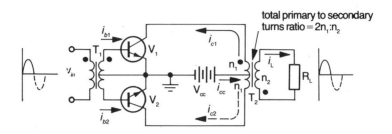

Fig 4-4 Transformer-coupled push-pull amplifier

a In this case V_1 and V_2 are a matched pair of silicon transistors. Under quiescent conditions both are biased off, since both bases are held at ground potential via the secondary winding of transformer T_1.

b Transformer T_1 splits the input signal (assumed sinusoidal, into two equal amplitude, antiphase signals to drive V_1 and V_2. During positive half cycles of the signal at the base of V_1, V_1 will conduct and V_2 will be reverse biased, and vice-versa during negative half cycles.

c It should be noted that i_{C_1} and i_{C_2} flow in opposite directions through their halves of the primary of transformer T_2, which results in the formation of a sinusoidal signal at the secondary of transformer T_2.

d Transformer T_2 not only acts to combine the signal from V_1 and V_2 into a sinusoidal signal at R_L, but also, if its turns-ratio is chosen correctly, optimises the reflected load resistance presented to V_1 and V_2 by maximising the power delivered to R_L. Note that the transformer T_2 normally does not match the load to the transistors in the maximum power transfer sense, rather the match is one which takes into account the limited voltage, current and power dissipation capabilities of the transistors.

e Several important relationships may be shown in the ideal case where all losses due to saturation are neglected:

    i $i_{CC} = i_{C_1} + i_{C_2}$
    ii $R'_L = n^2 R_L$
    iii $i_L = \dfrac{n_1}{n_2}(i_{C_1} - i_{C_2})$
    iv $P_{CC} = \dfrac{2V_{CC}^2}{\pi R'_L}$ at $P_o(\text{max})$ (See Section 4.26A for Proof.)

The respective waveforms are illustrated in Fig 4.5.

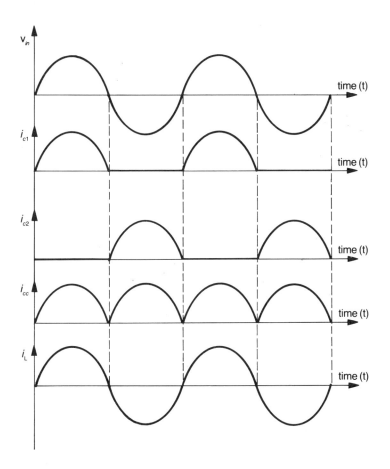

Fig 4-5 Waveforms for the push-pull amplifier circuit in Fig 4-4

4.6 Crossover distortion

The V_{BE}, I_C characteristic of an npn transistor is shown in Fig 4.6. It may be seen that the relationship between I_C and V_{BE} becomes markedly non-linear for V_{BE} values below the *knee*. This results in distortion of the output signal called **crossover distortion**.

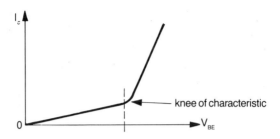

Fig 4-6 V_{BE}, I_c characteristic of npn transistor

The effect of non-linearity is shown in Fig 4.7. The non-linearity is introduced when the forward biasing input signal, which is effective in producing the output signal, crosses over from one transistor to the other.

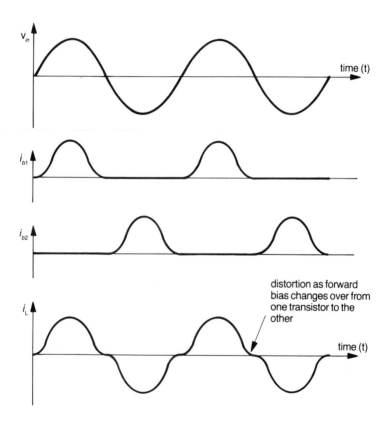

Fig 4-7 Illustration of the effect of crossover distortion

Clearly, if the input waveform signal is to be reproduced accurately in the output circuit, the crossover distortion must be considerably reduced. The basis of the usual method of minimising crossover distortion is as follows:
a   A small forward bias is applied to the base emitter junctions of V_1 and V_2. This places the operating points of V_1 and V_2 more or less at the knee of their respective V_{BE}, I_B characteristics.
b   The V_{BE}, I_C characteristics of the transistors V_1 and V_2 (Fig 4.4) now effectively become as shown in Fig 4.8.

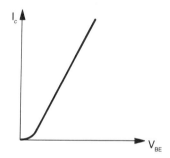

Fig 4·8 V_{BE}, I_c characteristic of forward biased npn transistors

A typical bias circuit is shown in Fig 4.9.

a   Diode V_3 provides sufficient voltage drop to bias V_1 and V_2 to the knee of their input volt-ampere characteristic. Resistor R determines the diode's bias current.
b   The diode also helps to stabilise the Q points against temperature variation, since its forward voltage drop, V_D, and the transistors' V_{BE}s decrease with temperature in the similar fashion necessary to maintain constant I_C.

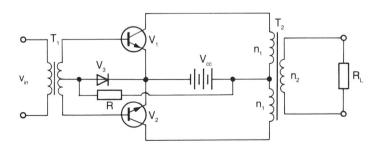

Fig 4-9 Transformer-coupled push-pull amplifier with forward biasing

4.7 Transistor power relationships for ideal push-pull amplifier circuit

Let $P_o(\text{max})$ = Maximum power output.

a The efficiency at $P_o(\text{max}) = 78.5\%$ (for proof see Section 4.26A). This implies that 78.5% of the supply power is converted into useful output power and 21.5% is dissipated in heating up the transistors, ie about 10.75% in each or about $\dfrac{P_o(\text{max})}{7\cdot3}$ watts

b It might be thought that the transistors would dissipate most power in themselves at $P_o(\text{max})$. This is not so. It can be shown that the transistors' average power dissipation is greatest when they are delivering approximately 40% of maximum power output to the load. At this point the efficiency is about 50% and the dissipation in each transistor is about $\dfrac{P_o(\text{max})}{5}$

c The peak instantaneous power dissipation in the transistor is $\dfrac{P_o(\text{max})}{2}$, but this is only of significance when operation occurs at low frequencies where the thermal time constant of the transistor is relatively short enough to allow the transistor to respond to the instantaneous power dissipation.

4.8 Disadvantages of a transformer as a phase splitter

a Distortion due to poor frequency response. The upper end of the frequency range is affected by the stray capacitance and inductance of the transformer. The lower end is affected by the primary self-inductance. The poor frequency response also produces considerable phase shift which makes the successful application of negative feedback difficult to achieve, because the feedback may turn out to be positive at some frequencies.

b The non-linear characteristic of the magnetic core material of the transformer.

4.9 Complementary symmetry amplifier

The push-pull circuit was originally a legacy from thermionic valve circuits, when only the 'npn type' valve, with conduction from plate to cathode, was available. With the advance of transistor design, both npn and pnp type transistors with similar characteristics became available and these enabled the transformer to be eliminated by using the now familiar complementary symmetry circuits.

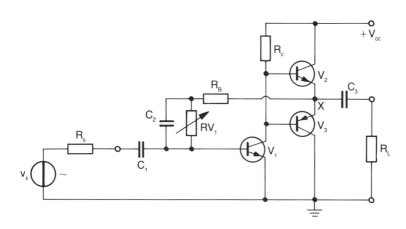

Fig 4·10 Basic circuit for a complementary symmetry amplifier

RV_1 sets the collector current of V_3 and so by varying the voltage drop across R_C provides optimum bias at X for V_1 and V_2. V_3 is a common emitter stage, providing voltage gain for the amplifier of about $-\dfrac{R_B}{R_S}$ due to the negative feedback provided by R_B. Readers should consult other appropriate texts for details of complementary symmetry amplifiers.

4.10 Development of a power amplifier using an operational amplifier

A development of a complementary symmetry amplifier could be a circuit as shown in Fig 4.11. The power output capability of the monolithic IC device is boosted by the addition of the complementary symmetry output stage composed of V_1 and V_2.

Operational Amplifier Circuits

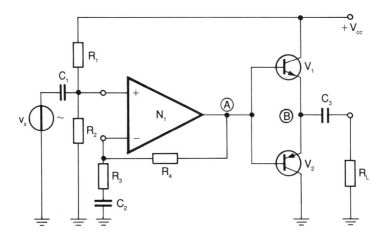

Fig 4·11 Power amplifier utilising an operational amplifier voltage gain stage

a   When $R_1 = R_2$, the quiescent voltage at the inverting terminal of N_1 will be $\dfrac{V_{CC}}{2}$.

b   'Earthing' R_3 via C_2 rather than directly to ground provides two benefits:

   i   The gain of the circuit at medium frequencies is $1 + \dfrac{R_4}{R_3}$. This reduces in value at low frequencies, as the increasing reactance of C_2 increases the impedance of the R_3 branch of the circuit. This is useful in audio circuits where undesirable signals may be present at much lower frequencies than the desired signal.

   ii At DC, the circuit from the non-inverting input of N_1 to its output becomes a voltage follower due to C_2 becoming an open circuit element. This forces the quiescent bias level at point A to be $\dfrac{V_{CC}}{2}$, thereby enabling maximum obtainable power output to be realisable when required.

c   By moving the point A connection of R_4 to point B, crossover distortion is considerably reduced since the source of the distortion is then placed within the feedback loop where it is reduced by the same factor that the other properties are improved by negative feedback.

Additional notes on biasing are given in Section 4.26B.

4.11 Bridge amplifiers

The power available from the various power amplifier circuits is proportioned to $\dfrac{V_{CC}^2}{R_L}$, hence any increase in output power is only possible by increasing V_{CC} or decreasing R_L. Fig 4.12 represents two power amplifiers connected in bridge configuration. This arrangement provides a large increase in maximum output power capability over that available from a single amplifier, but with no increase in supply voltage or change in R_L. This is especially useful when the supply rail voltage has a fixed predetermined value, as in the case of automobile audio equipment.

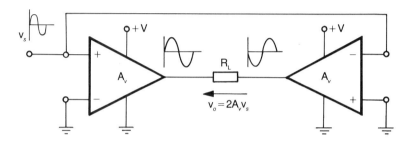

Fig 4-12 Bridge amplifier

In this arrangement, two amplifiers with numerically equal voltage gains, A_V, are used with anti-phase outputs, ie one inverting and one non-inverting amplifier. If the same signal is applied to both amplifier inputs, the magnitude of the output signal across the load will be twice that obtainable for a single amplifier, providing the load, R_L, is connected between the two amplifier outputs.

Thus, if it is within the two amplifier capabilities, the power output will be four times that from a single amplifier.

Example 1

In the bridge amplifier basic circuit diagram shown in Fig 4.13 it is assumed that points A and B are biased to a quiescent value of 20 V and that the power amplifiers N_1 and N_2 are operating ideally with no losses. Determine the efficiency of the bridge amplifier at (a) maximum power output, and at (b) 10 watts output. Assume that the integrated power amplifiers are ideal, ie they experience no losses due to saturation and both have a complementary symmetry output configuration.

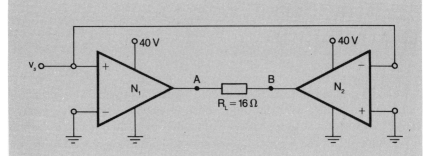

Fig 4-13

Solution

At maximum power output (P_o (max))

The maximum output signals at points A and B will be as shown in Fig 4.14.

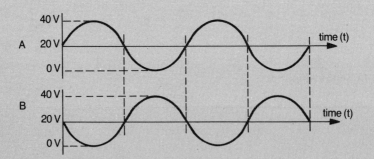

Fig 4-14 Waveforms at A and B at P_o (max)

The supply currents to N_1 and N_2 would be as shown in Fig 4.15.

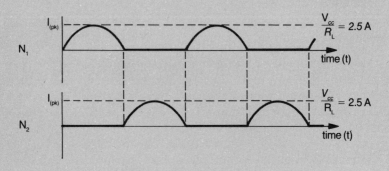

Fig 4-15 N_1 and N_2 supply current waveforms at P_o (max)

The power drawn from the supply by each amplifier would have an average value of: $40 \times I_{av}$ W = $40 \times \dfrac{2.5}{\pi}$ W = $\dfrac{100}{\pi}$ W (where I_{AV} = average supply current).

Hence, total supply power = $\dfrac{200}{\pi}$ watts for the two amplifiers.

The efficiency at $P_o(\max)$ = $\dfrac{P_o(\max)}{\text{total supply power}} \times 100\%$

= $\dfrac{50}{200/\pi} \times 100\% = 78.5\%$

At 10 W power output

The load current (rms) value would be given by: $I_{(rms)}^2 R_L = P_o$

so $I_{(rms)}^2 \times 16 = 10$. $I_{(rms)} = \sqrt{\dfrac{10}{16}} = 0.79$ A

so $I_{(0\text{-pk})} = 0.79 \times \sqrt{2} = 1.118$ A

Hence: $I_{DC_{(av)}} = \dfrac{I_{(0\text{-pk})}}{\pi} = 0.3559$ A

and $P_{CC} = V_{CC} \times I_{DC_{(av)}} = 40 \times 0.3559 \times 2 = 28.46$ W

The efficiency = $\dfrac{10 \text{ W}}{28.46 \text{ W}} \times 100\% = 35.14\%$

Example 2

In the circuit diagram shown in Fig 4.16, the amplifiers (N_1 and N_2) each have a gain of 26 dB. The respective outputs are automatically biased to $\left(\dfrac{V_{CC}}{2} \pm 1\right)$ volts and they saturate when their output voltages are within 2 volts of the supply rails.

Determine: (a) the variation in maximum output power due to the variation in output bias voltage settings, and (b) the range of output power available corresponding to variation of R_{v_1}, if $v_s = 1$ volt (rms) and the bias voltages are both set to 9 V.

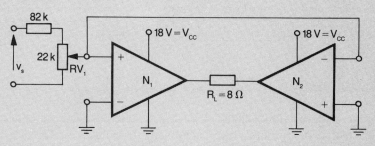

Fig 4·16

Solution

a Maximum power, $P_o(\text{max})$, capability without clipping will be attained when output bias setting is exactly 9 V for both amplifiers. Then, allowing for a total of 4 volts loss due to saturation voltages, the peak swing across the load will be $(18 - 4) = 14$ V.

so $P_o(\text{max}) = \dfrac{V_{rms}^2}{R_L} = \dfrac{(14/\sqrt{2})^2}{8} = 12 \cdot 25$ W.

The minimum value of $P_o(\text{max})$ will occur when at least one of the amplifier's bias voltages is at an extreme value, such as 8 or 10 volts. The maximum unclipped (0-pk) swing at the amplifier's output would then be limited to 6 volts (allowing for 2 volts saturation level). This would give a (0-pk) swing across the load of 12 volts maximum, without clipping.

Then minimum $P_o(\text{max}) = \dfrac{V_{rms}^2}{R_L} = \dfrac{(12/\sqrt{2})^2}{8} = 9$ W

Hence variation in maximum output power is from 9 to 12·25 W = 3·25 W.

b V_{in} will vary from: $\left(\dfrac{22\text{ k}}{82\text{ k} + 22\text{ k}}\right) 1000$ mV$_{(rms)}$ to zero V

$= 211 \cdot 5$ mV to zero V.

Hence, the rms signal at each amplifier output $= 211 \cdot 5$ mV $\times$ voltage gain. Voltage gain of each amplifier, A_v, is such that $20 \log_{10} A_v = 26$ dB

so $\log_{10} A_v = 1 \cdot 3$, hence $A_v = 20$ times,

ie rms signal output from each amplifier $= 211 \cdot 5$ mV $\times 20$ to zero
$= 4 \cdot 231$ volts (rms) to zero

The (0-pk) equivalent of $4 \cdot 231$ volts (rms) $= 5 \cdot 983$ V.

This gives a (0-pk) load voltage swing maximum of $2 \times 5 \cdot 983 = 11 \cdot 97$ V.

Hence, maximum output power would be: $\dfrac{(11 \cdot 97/\sqrt{2})^2}{8} = 8 \cdot 95$ W

so range is from $8 \cdot 95$ W to zero.

4.12 Integrated circuit power amplifiers

The general design philosophy of integrated circuit power amplifiers is similar to that of discrete power amplifiers. It would therefore, at this stage, be worthwhile considering some important design factors in discrete power amplifiers.

a The input stages combine to give high voltage gain, whilst the output stage gives unity voltage gain, but high current and hence power gain.
b High maximum slew rate capabilities are required because of the large voltage swing requirements. Consider a 50 V (pk-pk) signal that is required at 18 kHz, then $S \geq 2\pi f\, V_p = 2\pi \times 18 \times 10^3 \times 25$ V/s = 2·83 V/μs is necessary. The design factors needed for high slew rate to be achieved are beyond the level of this present text, however Section 4.13 examines a few important factors pertaining to discrete power output stages.

4.13 Discrete power output stage

Fig 4.17 represents a very much simplified discrete power output stage.

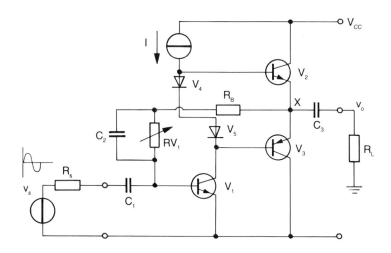

Fig 4·17 Sample discrete power output stage

a R_{V_1} is used to adjust the base current of V_1. This, in turn, varies the collector current of V_1 and the quiescent voltages at the bases of V_2 and V_3. The overall result is that the voltage at point X is varied and adjusted to approximately $\dfrac{V_{CC}}{2}$ to obtain equal maximum positive and negative voltage swing capabilities across R_L. This ensures that the circuit is adjusted to a point such that maximum output power capability, without distortion due to clipping, is achieved.
b V_1 provides voltage gain, whilst V_2 and V_3 are emitter followers giving unity voltage gain but considerable current gain.

Operational Amplifier Circuits

c R_B and R_{V_1} provide DC shunt feedback to stabilise the DC operating voltage at point X.

d R_B and the driving source impedance R_S provide feedback at signal frequencies to give a gain close to $-\dfrac{R_B}{R_S}$, in a similar way to the inverting amplifier circuit.

e The collector current of V_1, derived from the current source, causes a voltage drop across diodes V_4 and V_5 to provide the forward bias necessary to minimise crossover distortion in V_2 and V_3.

f The current source itself, because of its high internal impedance, ensures that the voltage gain of V_1 is quite large.

4.14 Bridge amplifier circuit incorporating IC power amplifiers

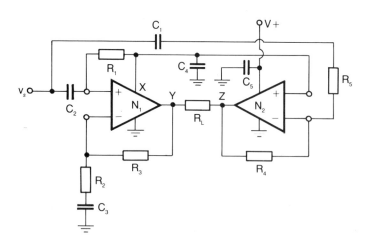

Fig 4-18 Bridge amplifier using integrated circuit power bridge amplifier incorporating IC power amplifiers

a The load R_L is connected between the output of an inverting amplifier N_2 and a non-inverting amplifier N_1.

b The gains of the two amplifiers are made equal and stabilised by the feedback presented by R_2, R_3 and R_4, R_5.

c The gain of the inverting stage at signal is $-\dfrac{R_4}{R_5}$ while that of the non-inverting stage is $1 + \dfrac{R_3}{R_2}$. Hence $1 + \dfrac{R_3}{R_2} = \dfrac{R_4}{R_5}$ for equal gains.

d The terminal X provides a voltage which is equal to half the supply voltage. This is fed to N_1 via resistor R_1. The resistor R_1 serves to isolate the terminal X from the input signal to avoid loading effects. Since loading effects are not present in the case of the inverting amplifier, no isolating resistor is needed for the connection to the non-inverting input of N_2.

e Resistors R_3 and R_4 combine with capacitors C_3 and C_1 to make N_1 and N_2 function as DC voltage followers between their non-inverting inputs and outputs, forcing the quiescent voltages at outputs Y and Z to be at $V_X = \dfrac{V^+}{2}$ supply volts.

f Capacitor C_4 acts to 'clean up' any unwanted signals that might be present on the bias line, while C_5 performs the same function for the main supply rail.

4.15 Basic heat sink theory

a When electrical power is dissipated within a semiconductor device it is converted into heat.

b Since the semiconductor junction would initially have been at the same temperature as its immediate surroundings, the ambient temperature, the junction will increase its temperature above its surroundings to an extent determined by the electrical power dissipated and by its ability to lose heat to its surroundings.

c The term 'thermal resistance' is nothing more than an indication of how difficult it is for heat to pass through the material in question to some point at a lower temperature. The original electrical power dissipated is converted into thermal energy, which may be thought of as escaping to lower temperature surroundings in the form of a *thermal current* which obeys a thermal 'Ohm's law'.

d Thermal resistance $= \dfrac{\text{temperature drop}}{\text{power dissipated}}$ and so has the physical dimension of °C/W. This text uses the symbol θ_{a-b} for the thermal resistance between points a and b of a thermal circuit.

e Just as an electrical circuit may have resistances in series or in parallel, so may a thermal circuit. The resistance from semiconductor junction to ambient is the series combination of the thermal resistance from junction to case and case to ambient, ie $\theta_{j-a} = \theta_{j-c} + \theta_{c-a}$.

f Often the junction of the device is not able to cool via the relevant thermal resistance path to its surroundings at a sufficient rate to prevent overheating occurring. It is then necessary to reduce this thermal resistance by using a heat sink.

g The heat sink must have a low thermal resistance between itself and the case, ie it must make good thermal contact with the device case. It must

Operational Amplifier Circuits

also have a low thermal resistance between itself and the air surrounding the device. Therefore, the device is often bolted to a heat sink which has a large surface area in contact with the air.

h A matt black finish and suitably shaped fins give good radiation, conduction and convection characteristics from the heat sink to its surroundings.

i In extreme situations, forced air cooling by means of a fan may also be necessary.

4.16 Heat flow model

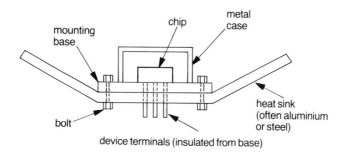

Fig 4-19 Mechanical diagram of a semiconductor device mounted on a heat sink

Most of the heat generated inside the device is transferred to the relatively heavy mounting base and then to the case, from which it is dissipated to the surroundings. In general, the case is not large enough to adequately dissipate the generated heat. Hence, the effective surface area must be increased by attaching a heat sink.

There are typically *three stages* to the heat transfer system of a semiconductor heat sink combination, each having its own thermal resistance to the next stage. The equivalent thermal diagram is shown in Fig 4.20.

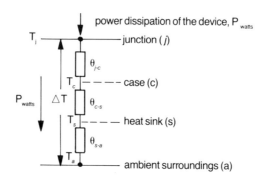

Fig 4-20 Thermal resistances

Power amplifiers and power supplies

A summary of the relationships existing in the thermal equivalent circuit is as follows:

a $\Delta T = P\theta_{j-a}$
where $\Delta T = T_j - T_a$ = (Junction temperature − ambient temperature)
and θ_{j-a} = Thermal resistance from junction to ambient in °C/W.
P = power dissipated at device function in watts.

b $\theta_{j-a} = \theta_{j-c} + \theta_{c-s} + \theta_{s-a}$
where θ_{j-c} = Thermal resistance from junction to case,
θ_{c-s} = Thermal resistance from case to heat sink,
θ_{s-a} = Thermal resistance from heat sink to ambient.

c It should be noted that parallel thermal paths, such as that directly from case to ambient, normally have thermal resistances much higher than the alternative path via the heat sink and so can be neglected, ie $\theta_{c-a} \gg \theta_{c-s} + \theta_{s-a}$.

Example 3

A power amplifier's output transistors, mounted on the same heat sink, dissipate a maximum of 40 watts each. The following information is supplied regarding the transistors:

a Operating temperature range = −65°C to +200°C
b Thermal resistance, junction to case = 0·885°C/W
c Thermal resistance, case to heatsink = 0·33°C/W

Refer to Fig 4.21, and determine the maximum permissible thermal resistance of the heat sink for an ambient temperature of 35°C

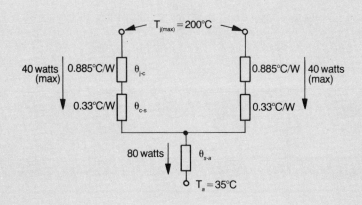

Fig 4-21

Solution

Let $T_j \, max$ = maximum permissible junction temperature. The life expectancy and reliability of the device are dependent on this not being exceeded.

Then $T_a + 2P\theta_{s-a} + P(\theta_{c-s} + \theta_{j-c}) \leq T_j \, max$

ie $35 + 80\theta_{s-a} + 40(0{\cdot}33 + 0{\cdot}885) \leq 200$

$\theta_{s-a} \leq 1{\cdot}455°C/W$.

4.17 Introduction to regulators

A voltage regulator is intended to provide a constant (almost constant, in practice) voltage across a load in spite of variations in the supply voltage and in the load current requirements. The variations in the supply voltage may be variations in the level of the DC supply itself, such as would occur if the supply was derived from the AC mains source of fluctuating level, or could be an AC ripple on such a supply, or both types of variations could be present.

Line regulation is the change in output voltage, for a specified change in input voltage, usually expressed as a percentage.

Load regulation is the change in output voltage, for a specified change in load current, often expressed as a percentage.

4.18 Basic discrete series regulator circuit

Fig 4.22 represents a basic discrete series regulator circuit from which it can be seen that: $V_L = V_Z - V_{BE}$.

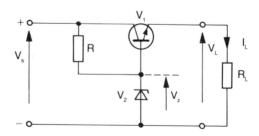

Fig 4-22 Basic discrete series regulator circuit

a It can be seen from $V_L = V_Z - V_{BE}$ that, should V_L attempt to decrease, then V_{BE} will, as a consequence, increase. (V_Z remains virtually constant). This will result in I_L increasing, hence increasing V_L in opposition to the original attempted change.

b This is a relatively insensitive circuit, since I_L responds only to the direct comparison of V_L with V_Z.

c In the more elaborate circuits used in high performance discrete regulators, the error signal $(V_L - V_Z)$ is amplified and used to control V_L.

4.19 Basic regulator with increased loop gain

In Fig 4.22 the zener diode V_Z is the reference element which determines the stability of the output voltage. In Fig 4.23 a predetermined fraction of the output voltage is compared with V_Z. The fraction is governed by resistors R_1 and R_2.

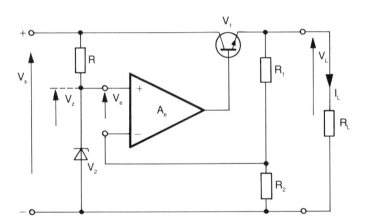

Fig 4-23 Basic regulator with increased loop gain

a If the output voltage is such that $\left(\dfrac{R_2}{R_1 + R_2}\right)V_L$ is greater than V_Z, then the error amplifier, A_e, will have a negative error signal, V_e, driving it.

This will cause the output of A_e to decrease the voltage applied to the series pass transistor, V_1, which, by virtue of V_1 being in common collector configuration, will cause V_L to decrease. Conversely, V_L will tend to increase if $\left(\dfrac{R_2}{R_1 + R_2}\right)V_L$ is less than V_Z.

Operational Amplifier Circuits

b This implies that any deviation from $\left(\dfrac{R_2}{R_1 + R_2}\right) V_L$ equalling V_Z will be opposed by the consequent changes in signal level at the base of V_1.

c In Fig 4.23 equilibrium will be reached when:

$$V_Z = \left(\dfrac{R_2}{R_1 + R_2}\right) V_L.$$

One of the advantages of using the resistors R_1 and R_2 to divide the output voltage before comparing it with V_Z, is that V_L is no longer directly dependent on V_Z, but can be adjusted to some required level or varied over a wide range of voltages subject to the overall condition that V_L must be greater than V_Z.

If V_Z is a 4·7 V zener diode, biased via R, and if $R_1 = 10$ k and $R_2 = 4\cdot7$ k then V_L may be determined as follows:

$$V_L = \left(\dfrac{R_1 + R_2}{R_2}\right) V_Z, \text{ ie } V_L = \left(\dfrac{14\cdot7}{4\cdot7}\right) 4\cdot7 = 14\cdot7 \text{ volts}$$

To obtain 14·7 volts from the basic regulator circuit shown in Figure 4.23 would require a 15·3 volt zener diode, or thereabouts. Since the closest standard value is 15 volts this would cause problems, especially if many regulators were to be constructed, as for a production run. If a potentiometer is incorporated in the divider chain, as in Fig 4.24, then the circuit could have its output adjusted to 14·7 volts for any specific zener diode of a type specified as having a zener voltage of 4·7 ± 10% volts, at the appropriate bias current level.

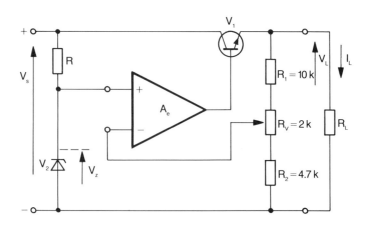

Fig 4-24 Regulator circuit with adjustable output voltage

Example 4

The circuit shown in Fig 4.25 is a voltage regulator giving an output voltage of 15 volts.

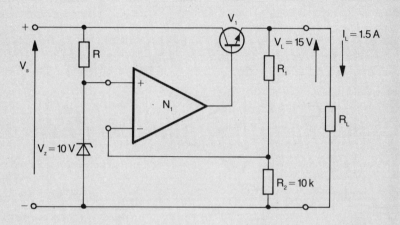

Fig 4·25

Determine **a** the value of R_1,
b the power dissipated in the series pass transistor V_1,
for an unregulated input of 24 V.

Solution

a $V_Z = \left(\dfrac{R_2}{R_1 + R_2}\right) V_L$ at equilibrium, ie $\dfrac{R_1 + 10 \text{ k}}{10 \text{ k}} = \dfrac{15}{10}$

so, $R_1 = 5$ k.

b Power dissipated $= V_{CE} I_{CE}$
$= (V_S - V_L) I_L$, neglecting current in R_1
$= (24 - 15) 1·5$
$= 13·5$ watts

4.20 Regulators producing output voltages lower than the internal reference element

In general, the regulators previously discussed inherently produce output voltages greater than V_{ref} since, with reference to Fig 4.23,

Operational Amplifier Circuits

$$V_L = \left(\frac{R_1 + R_2}{R_2}\right)V_Z = \left(1 + \frac{R_1}{R_2}\right)V_Z.$$

To obtain voltages lower than V_Z involves only a simple change to the circuit as shown in Fig 4.26.

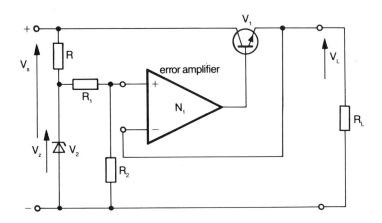

Fig 4-26 Basic circuit for regulators producing output voltages lower than the internal reference element

The circuit shown in Fig 4.26 supplies a *divided* reference voltage to the non-inverting input of the error amplifier and has a direct connection from the output to the inverting input of that amplifier.

Equilibrium will now come about when:

$$V_L = \left(\frac{R_2}{R_1 + R_2}\right)V_Z,$$

ie when V_L has a value determined by R_1, R_2 and V_Z but is inherently $\leq V_Z$.

4.21 Pre-regulation

The output voltage of a regulator varies in a manner proportional to the voltage of the reference element. It follows that a precision regulator requires a highly stable reference element voltage. Referring to Fig 4.26, relevant details of circuit action are as follows:

a The variations in supply voltage produce significant variations in the voltage, V_Z, across the zener diode V_2.

Power amplifiers and power supplies

b The current in the zener diode, V_Z, is given by: $\dfrac{V_S - V_Z}{R}$, hence if V_S varies from V_{S_1} to V_{S_2}, the current in the diode will increase from $\dfrac{V_{S_1} - V_Z}{R}$ to $\dfrac{V_{S_2} - V_Z}{R}$; ie an increase of $\dfrac{V_{S_2} - V_{S_1}}{R}$ amperes.

c It is assumed in (b) that the change in V_Z, although significant as far as its reference element properties are concerned, is small by comparison with changes in V_S.

d If the zener diode has an average dynamic resistance, R_Z, the change in V_Z will be $\left(\dfrac{V_{S_2} - V_{S_1}}{R}\right) R_Z$ volts.

e Let V_{S_1} = 18 volts, V_{S_2} = 24 volts and V_Z is a 10 V zener diode with R_Z = 8·5 ohms. R = 910 ohms, giving a nominal bias current of 12 mA for V_S = 21 V. The change in V_Z, due to changes in V_S = $\left(\dfrac{24 - 18}{910}\right) 8\cdot 5$ = 56 mV, which is of considerable significance when designing a precision regulator.

The problem of variation in V_Z with V_S can largely be overcome by driving the reference element from a constant current source as shown in Fig. 4.27.

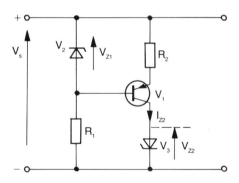

Fig 4-27 Pre-regulation of reference element voltage

a The current biasing V_3 is given by: $\dfrac{V_{Z_1} - V_{BE}}{R_2}$.

b V_{Z_1} will vary slightly with V_S and V_{BE} will vary slightly with changes in I_{Z_2}. These small changes will only produce small variations in I_{Z_2}, and the subsequent changes in V_{Z_2} will be much smaller than those without additional pre-regulator circuit.

4.22 Current limiting circuit

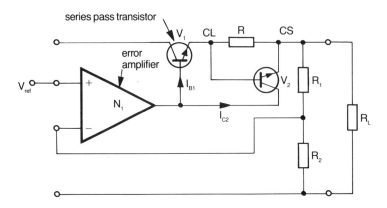

Fig 4-28 Current limiting circuit

a The current limiting circuit (Fig 4.28) consists of V_2 and R.
b Neglecting the current flowing in R_1 and R_2 as being relatively much smaller, the load current, I_L, flows through R and, if it increases to a value where $I_L R \simeq 0.65$ V, will forward bias transistor V_2.
c This will cause collector current I_{C_2} to flow to the load via V_2 instead of acting as part of the base current I_{B_1} of V_1.
d Clearly, every milliamp of current that flows as I_{B_1} is converted to β_1 milliamps of load current by V_1, whereas if it is diverted via V_2 it only contributes 1 mA to the load current.
e Since the effect of introducing this circuitry is to limit the load current to a value only slightly more than the value that produces 065 volts drop across R, it is a current limiting circuit.
f The connection from R to the base of V_2 is traditionally called the **current limit connection**, CL, whilst the connection from the emitter of V_2 to R is called the **current sensing connection**, CS.

4.23 Integrated circuit regulators

There are two basic types of integrated circuit regulators:
a The three terminal regulator which incorporates all the basic elements of regulators previously mentioned in this text, and usually a thermal overload circuit as well. They give a pre-determined output voltage.
b The type of regulator in which everything is incorporated, except the

actual voltage divider network, which establishes the regulated voltage, and the current limiting resistor.

4.24 Three terminal regulators

Fig 4.29 represents the basic concept of a three terminal regulator.

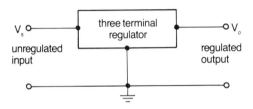

Fig 4-29 Basic connection diagram for a three terminal regulator

Three terminal regulators are usually mounted directly on the circuit board, and are normally used where the board is largely, or entirely, made up of circuitry operating from a standard three terminal regulator voltage such as 5 V, 12 V, 15 V. Their use largely overcomes many of the interactive supply problems associated with the provision of one voltage level from a separate regulator to a number of boards requiring that voltage.

4.25 Integrated circuit series voltage regulator

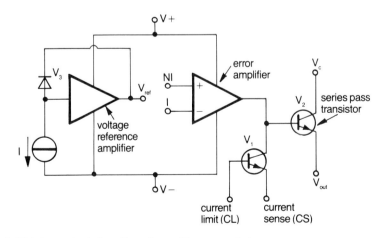

Fig 4-30 Integrated circuit series voltage regulator

Operational Amplifier Circuits

The circuit shown in Fig 4.30 does not show connections made for the purposes of preventing unwanted oscillation or for the production of negative output voltage regulators. Such connections are adequately described in manufacturers' literature.

Fig 4.31 shows a block diagram for a voltage regulator to give a regulated output voltage of about 5 V from an IC regulator with a nominal V_{ref} of 7·15 V.

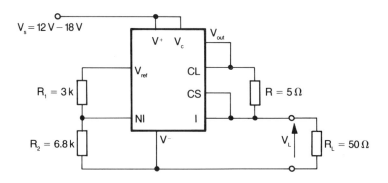

Fig 4-31 Basic block diagram of a regulator to produce $V_L < V_{ref}$ using an IC regulator device

a R_1 and R_2 represent the divider that feeds a reference voltage, derived from V_{ref}, to the non-inverting input of the error amplifier.

b   This voltage will be: $\left(\dfrac{R_2}{R_1 + R_2}\right) V_{ref} = \left(\dfrac{6\cdot 8\ k}{3\ k + 6\cdot 8\ k}\right) 7\cdot 15\ V,\ =$ 4·96 V (nominal value, since V_{ref} can vary slightly from unit to unit).

c   The regulated output voltage V_L is fed to the inverting input of the error amplifier, so equilibrium will occur when $V_L = 4\cdot 96$ volts.

d   R is the current sensing resistor and, if it is given in the manufacturer's data that the base emitter voltage of the internal current sensing transistor is, say, 0·65 V, then current limiting will occur when:

$I_L R = 0\cdot 65$ V, ie $I_L = \dfrac{0\cdot 65}{5} A = 130$ mA.

This is, of course, higher than the actual load current which would be about: $\dfrac{5\ V}{50\ \Omega} = 100$ mA.

e   The unregulated input voltage V_S is shown as varying between 12 and 18 V. In practice most integrated circuit regulators would be capable of handling values of V_S down to about 8 V for a 5 V regulated output.

Power amplifiers and power supplies

There must normally be a difference of at least 3 V between V_S and V_L to ensure that the series pass transistor does not go into saturation. It is essential that this differential voltage is maintained when there is ripple on the input supply line.

4.26 Important review points

A Proof of:

a $P_{CC} = \dfrac{2V_{CC}^2}{\pi R_2}$ at P_o (max)   (From Section 4.5.)

b the efficiency of P_o (max) = 78.5%   (From Section 4.7.)

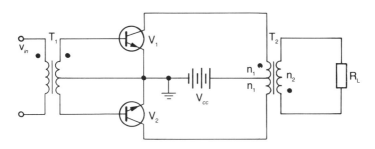

Fig 4-32 Transformer-coupled push-pull amplifier

a Average power from supply, $P_{CC} = V_{CC} \times I_{DC(av)}$ where $I_{DC(av)}$ is the average supply current. The peak supply current when P_o (max) is being delivered to the load is $\dfrac{V_{CC}}{R'_L}$, and since the supply current has the nature of a full wave rectified sinusoidal signal, its average value will be $\dfrac{2}{\pi} \dfrac{V_{CC}}{R'_L}$, so average supply power, $P_{CC(av)}$

$= V_{CC} \times \dfrac{2}{\pi} \dfrac{V_{CC}}{R'_L} = \dfrac{2}{\pi} \dfrac{V_{CC}^2}{R'_L}$.

Notes: **i** At less than P_o (max) P_{CC} equals $\dfrac{V_{CC} \times 2\, V}{\pi R'_L}$ .... (1) where V is the corresponding peak swing across each half of the primary of T_2; **ii** P_o then equals $\dfrac{V^2}{2R'_L}$ watts. **iii** Efficiency, in general is:

$\dfrac{V^2/2R'_L}{2\, V_{CC} \times V/\pi R'_L} \times 100\% = \dfrac{25\pi V}{V_{CC}}\%$.

105

b The peak sinusoidal voltages swing across the load is V_{CC}, so:

$$P_o(\text{max}) = \frac{(V_{rms})^2}{R'_L} = \frac{\left(\frac{V_{CC}}{\sqrt{2}}\right)^2}{R'_L} = \frac{V^2_{CC}}{2R'_L}.$$

From (1): $P_{CC} = \dfrac{2 V^2_{CC}}{\pi R'_L}$ at P_o (max)

so efficiency $= \dfrac{\dfrac{V^2_{CC}}{2R'_L}}{\dfrac{2V^2_{CC}}{\pi R'_L}} = \dfrac{\pi}{4} \times 100\%$

$= 78 \cdot 5\%$ at P_o (max).

B Notes on biasing for single supply operation

From Section 4.10.

In order that maximum output voltage swing may be obtained, with sinusoidal operation, it is necessary to bias the op-amp output to $\dfrac{V_{CC}}{2}$ volts where V_{CC} is the single bias rail voltage.

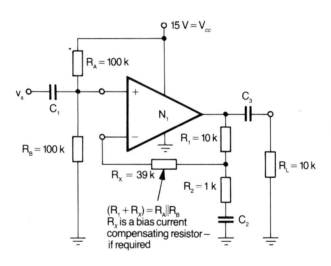

Fig 4·33 Single supply biasing arrangements

Power amplifiers and power supplies

The arrangement in Fig 4.33 will accept sinusoidal input signals, amplify them 11 times at frequencies where the reactance of capacitors may be neglected, and produce a sinusoidal output voltage. C_1 both prevents DC entering the signal source and also prevents the signal source shorting out the DC voltage across R_B. C_2 causes the circuit to act as a voltage follower at DC from non-inverting input to output since it does not allow DC to pass through R_2. This means that feedback is via R_1 and R_X. Since (+) input is at 7·5 volts so must the (−) input and output also will be at 7·5 volts under quiescent conditions. C_3 prevents the DC bias voltage at the output passing to the load so the actual output across R_L is a sinusoidal signal symmetrical about zero volts.

C Alternative biasing arrangement requiring one less capacitor

The circuit shown in Fig 4.34 has a gain of 11 at operational frequencies which falls to zero at DC. Since the gain from the + input to the amplifier output remains at 11 down to DC this means that any bias voltage present at the + input will also be amplified with this gain.

For the gain to stay at 11 down towards DC means that any bias voltage present at the (+) input will also be amplified by the gain of 11 (Fig 4.34).

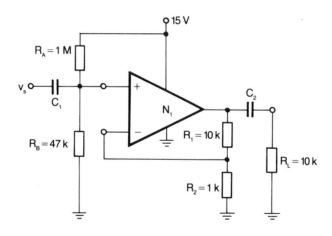

Fig 4·34 Alternative single supply arrangements

To obtain an output bias level of 7·5 V would require that the bias level at the (+) input be lowered to (7·5/11) volts, ie 0·682 V. This means that:

$$\left(\frac{R_B}{R_A + R_B}\right) V_{CC} = 0.682; \text{ ie } \frac{R_B}{R_A + R_B} = \frac{0.682}{15} = 0.0455.$$

hence $R_A = 21 R_B$.

Operational Amplifier Circuits

$R_A = 1$ M and $R_B = 47$ k would give a bias voltage of 0·673 V at (+) input and 7·41 V at output. The input impedance of the circuit at operational frequencies would now be about 45 k instead of the previous 50 k (ie $R_A \| R_B$) refer to Fig 4.33. Providing C_1 and C_2 were large capacity capacitors, the frequency response would extend down towards zero Hz, eg for input and output breakpoint at 1 Hz, one would require:

$$1 \text{ Hz} = \frac{1}{2\pi C_1 (R_A \| R_B)} \quad \text{hence } C_1 = 3\cdot55 \ \mu F$$

$$1 \text{ Hz} = \frac{1}{2\pi C_2 R_L} \quad \text{so } C_2 = 15\cdot9 \ \mu F.$$

4.27 Problems for readers

Example 5

Fig 4.35 shows the essential elements of a bridge power amplifier driving an 8 Ω load.

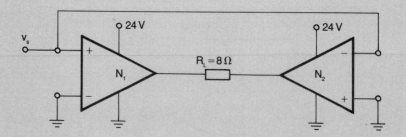

Fig 4-35

Each amplifier has a voltage gain of 30 times, has its output biased to 12 V, and saturates at the 22 V and 2 V output levels.
a Determine the value of v_S (rms) if the power delivered to the load is 2 watts.
b Determine the maximum power that can be delivered to the load and the value of v_S necessary to produce it.

Solutions

a 66·7 mV (rms)
b 25 watts, 236 mV (rms)

Example 6

In the circuit shown in Fig 4.36 it can be assumed that the capacitors have negligible reactance at operational frequencies.

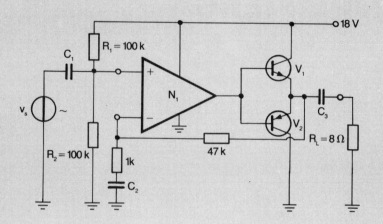

Fig 4·36

a Determine the maximum power output that can be delivered to the load if the saturation voltages of N_1, V_1 and V_2 are neglected.
b If N_1 saturates at the 2 V and 16 V levels, and V_1 and V_2 have collector-emitter saturation voltages of 1·5 volts, determine the maximum output power that can be delivered to R_L if the base emitter voltage drops of V_1 and V_2 are neglected.
c What is the approximate value of v_S (rms) necessary to obtain full output power under the condition of (b) above.

Solutions

a 5·06 watts
b 3·06 watts
c 0·103 volts (rms)

Chapter 5
SELECTED APPLICATIONS OF OPERATIONAL AMPLIFIERS

5.1 Introduction

Clearly, the operational amplifier is a popular and extremely useful device which has a very wide range of applications. Operational amplifiers are so versatile that their applications are limited only by the imagination of the circuit designer. Some uses of operational amplifiers are: alarm circuits, audio and electronic music circuits, automobile electronic circuits, basic building blocks in computers, many types of control circuits, analogue converters, digital work, medical electronics, displays, filters, video games, indicators, opto-electronics and photography, power supplies, switching/timing devices, test and measurement. Readers will find examples of such circuits in appropriate electronic magazines.

Design engineers take into consideration such factors as offset voltage, CMMR and slew rate when selecting an operational amplifier for use in a particular circuit, and they must make considerable use of the manufacturers' data sheets.

In many non-critical applications a general purpose operational amplifier such as the 741 type is usually adequate. However, there are many applications for which high frequency gain, very high slew rate or low input offset voltage (or some other special requirement) will require a special type of operational amplifier to give the desired performance.

Brief explanations will be given in this chapter of some operational amplifier applications in the following areas:
- audio circuits
- precision rectifiers
- high impedance DC voltmeter
- medical electronics
- measurement of incident radiation using photodiode
- full wave ideal rectifier

- peak detector
- variable gain AC amplifier
- differential light intensity circuit
- filter applications
- linear read-out amplifier for resistive bridge circuit
- bass and treble tone control circuit
- scratch filter circuit
- switching power amplifier
- log converter circuit
- musical chiming circuit
- an R-S flip flop

5.2 Audio circuits

Pre-emphasis Many audio circuits require carefully determined frequency responses. Pre-emphasising, or boosting the signal source's high frequency signal components and de-emphasising or reducing the response of the signal processing circuitry, is often used to improve the overall signal to noise ratios of audio systems whilst still retaining an overall flat frequency response. Operational amplifiers are well suited to these applications because of their high gain and easily manipulated closed loop frequency response.

Filters Filters are often used to improve the overall quality, and once again effective use of operational amplifiers can be made.

Tone control Tone control of audio involves modifying the flat response to emphasise low and/or high frequencies dependent upon listener preference. Operational amplifier circuits are able to satisfy these requirements most effectively.

Balance and loudness control Balance and loudness controls also use operational amplifiers. Due to the non-linearity of the human hearing system, low frequencies must be boosted at low listening levels. Balance, ie varying the relative level of the left and right-hand channel of stereo systems, level and loudness controls provide all the necessary listening controls to produce the desired audio/music response. Once again operational amplifier closed loop responses are readily tailored to meet such requirements.

Details of audio circuits may be found in appropriate texts.

5.3 Precision rectifiers

The ideal semiconductor diode would have infinite resistance when its anode was more negative than its cathode, and zero resistance when forward biased. In practice there is a threshold, or knee, voltage of about 0·6 volts, for silicon diodes, before significant conduction occurs when forward biased. This is a serious problem, in the case of low level

Operational Amplifier Circuits

applications of circuits such as rectifiers, clippers, clampers, AC voltmeters, rectangular to polar converters and peak detectors that utilize diodes. Fortunately operational amplifiers may be used in conjunction with the diodes to reduce the effective threshold voltage by a very large factor.

In the case of rectifier circuits the resulting circuits are termed either 'precision rectifiers' or 'absolute value circuits', depending on the application. Section 5.7 has details of a precision full wave rectifier circuit.

5.4 High impedance DC voltmeter

In Fig 5.1 the voltage to be measured, V_1, is applied to the (+) input terminal.

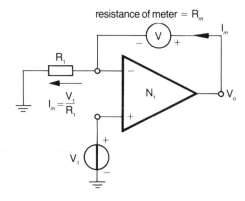

Fig 5·1 Basic high impedance DC voltmeter

a The meter current, I_m, is set by V_1 and R_1.
b V_1 sees the very high impedance of the (+) input. Since the (+) input draws negligible current, it will not load the voltage being measured.
c The meter resistance has no effect on the meter current.
d If the full scale deflection current for the meter is I_{fsd}, then to obtain full scale deflection requires $R_1 = \dfrac{V_1}{I_{fsd}}$.
e The actual input resistance seen by V_1 is given by:

$$R_{in}\left[\frac{A_{OL}}{A_{CL}}\right] = R_{in}\left[\frac{A_{OL}}{1 + \dfrac{R_m}{R_1}}\right]$$

where R_{in} is the input resistance of the operational amplifier and A_{OL} its open loop gain.

Example 1

A 50 μA *fsd* meter with internal resistance of 1 k is to be used in the circuit of a high impedance voltmeter intended to give full scale deflection when 1 volt is applied to it. Determine the voltmeter's input resistance and the value of R_1. Assume the operational amplifier's open loop gain and input resistance are 100 000 and 1 MΩ respectively.

For *fsd* with 1 V, need $R_1 = \dfrac{1\ \text{V}}{50\ \mu\text{A}} = 20\ \text{k}$

R_{in} seen by 1 V signal will be:

$$1 \times 10^6 \left[\dfrac{100\,000}{1 + \dfrac{1000}{20\,000}} \right] = 9\cdot 52 \times 10^{10}\ \Omega\ .$$

5.5 Application in medical electronic monitoring systems

Operational amplifiers are used effectively in many medical electronic monitoring systems. A common use (in the case of stroke patients) is monitoring the electromyogram (EMG) of the muscle.

When a muscle contracts, a very small voltage appears on the surface of the skin covering the muscle. The voltage waveform is called the EMG of the muscle. This EMG signal is usually less than 1 mV in amplitude and a differential amplifier with good common mode rejection ratio is required to amplify it to a useful amplitude, in isolation from the relatively noisy electronic environment. Such a function is appropriate to suitable operational amplifiers.

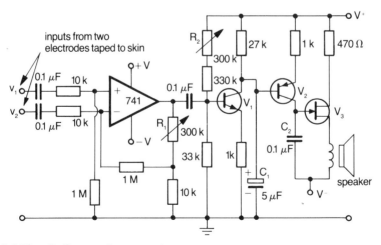

Fig 5-2 Circuit diagram for an audio EMG monitor

Operational Amplifier Circuits

a Fig 5.2 represents a schematic circuit diagram for an audio EMG monitor.
b The input is obtained from the two electrodes taped to the skin over the muscles being examined. The differential amplifier circuit gain can be varied by adjusting R_1.
c V_1 acts as a half wave rectifier by amplifying only half sections of the output from the operational amplifier. The bias of V_1 is controlled by R_2. C_1 averages the rectified output of V_1. Hence, the averaged amplified EMG signal controls the base current of V_2.
d V_2 controls the charging current of C_2 and hence the oscillation frequency of the unijunction oscillator, V_3, which in turn controls the pitch of the tone produced by the speaker.

5.6 Measurement of incident radiation using a photodiode

Fig 5.3 shows a photodiode in a circuit whereby the current I is converted into a voltage by the resistance, R.

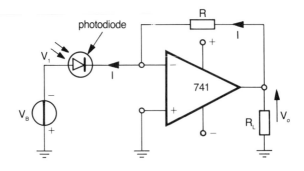

Fig 5-3 Measurement of incident radiation using a photodiode

a The photodiode is reverse-biased by V_B.
b In darkness, the photodiode possesses a small leakage current which is usually only a few nanoamperes in magnitude.
c Light striking the diode will normally increase the current to perhaps 50 μA or more. This current bears a linear relationship to the incident radiation over many decades of irradiance.

d The current depends only on the light striking the photodiode and not on V_B, so the output voltage, $V_o = -IR$ is proportional to I and hence to the radiation incident upon the photodiode.

5.7 Full wave ideal rectifier

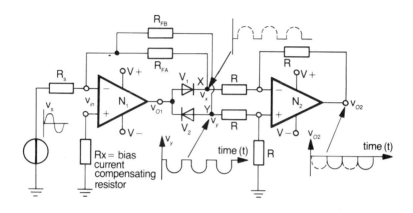

Fig 5·4 Full wave ideal rectifier

a Diodes V_1 and V_2 will start to conduct when the output of N_1 reaches a (0-pk) amplitude of about 0·6 V, because they are connected back to the inverting input of N_1 via R_{FA} and R_{FB}, and the inverting input will stay very close to ground potential due to the negative feedback.

b Assuming N_1 has an open loop gain of 200 000 V_1 and V_2 will conduct when v_{in} reaches about 3 µV (0-pk). Before this level is reached, N_1 is isolated from N_2 and the output, v_o, because V_1 and V_2 will be effectively open circuits to signals.

c There is no contribution to v_o as a result of v_{in} being applied directly to N_2 via R_{FA} and R_{FB} (providing $R_{FA} = R_{FB}$), because N_2 and the resistors R, R_{FA} and R_{FB} will act as a differential amplifier with zero differential input signal.

d V_1 conducts during negative half cycles of v_{in} and V_2 conducts during positive half cycles when the amplitude of v_{in} exceeds 3 µV. The current in $R_S = \dfrac{v_S}{R_S}$ because of the virtual earth at the (−) input. For practical purposes, all this current flows through either R_{FA} or R_{FB} depending upon which diode is conducting. It follows that the voltages at X and Y will be $-\left(\dfrac{R_F}{R_S}\right)v_S$, where $R_{FA} = R_{FB} = R_F$ with

Operational Amplifier Circuits

V_x being made up of positive half cycles and v_y negative half cycles.

e The resistors R, and N_2 act as a unity gain differential amplifier to v_y and v_x, so that:
$$v_o = (v_y - v_x).$$

f Since v_y consists of negative going half cycles and v_x of positive, v_o is made up entirely of negative going half cycles, as shown in Figure 5.4.

g The output is a full wave rectified version of v_S, without the errors introduced by the forward voltage drops of the diodes, as for conventional full wave or bridge rectifier circuits. Only 3 μV of the signal v_{in} is ineffective.

5.8 Peak detector

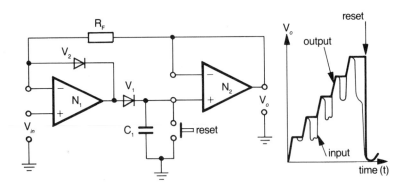

Fig 5-5 Peak detector

The peak detector is used in such areas as bar code readers and instrumentation circuits.

a A peak detector's output signal corresponds to the highest input voltage received during the period after the circuit has been reset.

b N_1 acts as a voltage follower to positive input signals since V_2 then conducts and acts as a negative feedback path.

c If v_{in} exceeds the voltage on C_1, then V_1 will conduct and allow C_1 to charge up to a voltage close to v_{in}. This voltage is transferred to v_o by N_2, which is connected as a voltage follower. The combination of N_1 and N_2 is also forced to act as a voltage follower because of the presence of R_F.

d If v_{in} is less than the voltage on C_1, then V_1 will not be forced into conduction.

e If V_2 was not part of the circuit, then N_1 would act as an open-loop gain device when v_{in} was less than v_o, and its output would be driven into saturation. There would then be a short period before the circuit could respond to any input voltage higher than that of C_1, because of the time taken to recover from the overdrive. V_2 prevents such saturation occurring, however, since it conducts if $v_1 < v_o$.

f The high input impedance of N_2 minimises the leakage from C_1, but not entirely, so there is a limit to the period for which v_o can be considered as corresponding to the highest input voltage.

g There is also a small reverse leakage via the reverse biased V_1. It follows that V_1 should be chosen to have a low value of reverse leakage current.

5.9 Variable gain AC amplifier

Variable gain AC amplifiers are used in such areas as audio circuits and instrumentation circuits.

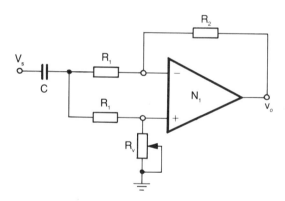

Fig 5-6 Variable gain AC amplifier

a R_V can be adjusted from zero to a maximum value of R_2.

b The voltage at the non-inverting input is $\left(\dfrac{R_V}{R_V + R_1}\right) v_S$, assuming C as having negligible reactance at operational frequencies.

c The negative feedback connection via R_2 ensures that the voltage at the inverting input is effectively that at the non-inverting input, so that nodal analysis gives the relationship:

$$\frac{v_S}{R_1} + \frac{v_o}{R_2} = \left(\frac{R_V}{R_V + R_1}\right) v_S \cdot \left(\frac{1}{R_1} + \frac{1}{R_2}\right)$$

Then, if R_V has its minimum value of zero, it can be shown that $v_o = -\left(\dfrac{R_2}{R_1}\right)v_S$

whereas if R_V has its maximum value of R_2, then it can be shown that $v_o = $ zero,

so gain is adjustable over the range zero to $-\dfrac{R_2}{R_1}$.

d The input resistance is given by: $R_1 \| (R_1 + R_V)$ so this varies from $\dfrac{R_1}{2}$ to $R_1 \| (R_1 + R_2)$. Consequently, if the maximum design gain was 100, the input resistance variation is from $0 \cdot 5\, R_1$ to $R_1 \| 101\, R_1$, ie $0 \cdot 99\, R_1$.

This is not a large variation, and for a low impedance source could be considered as more or less constant as far as its effect upon the circuit is concerned. For unity maximum gain the variation is from $0 \cdot 5$ to $0 \cdot 67\, R_1$.

5.10 Differential light intensity circuit

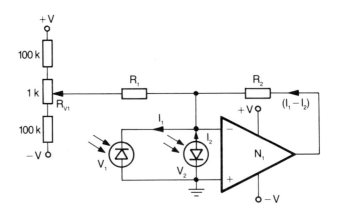

Fig 5·7 Circuit which responds to differential light intensity

a Fig 5.7 illustrates a circuit that could be used to provide a signal to a relay to, say, switch off electric lighting when the outside light exceeded that inside the room.

b The resistive divider circuit can be used to cancel out the effect of input offset voltage, by applying an opposing signal to the amplifier input by adjustment of RV_1.

c Since both ends of R_1 will ideally be at zero volts there will be no current in R_1. Hence, the combined current for the two photoelectric cells will flow via R_2, causing the amplifier output potential to be $(i_1 - i_2) R_2$ volts.

5.11 Filter applications

Fig 5.8 represents an example of a low pass filter.

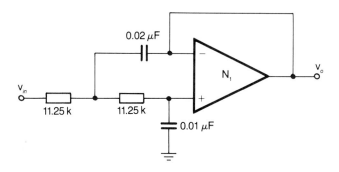

Fig 5-8 Low pass filter

a Operational amplifiers may be used to produce filters superior to those containing only passive elements.

b The main improvement is that the output impedance becomes very low, so that the load on the filter network has no effect upon its characteristics. High pass, low pass and band pass filters may be constructed.

c The circuit shown in Fig 5.7 has a -3 dB frequency of 1 kHz and is 40 dB down in its response at 10 kHz, and there is no transmission loss at low frequencies.

d If some gain is required at low frequencies, then the amplifier may be changed from a voltage follower to a non-inverting amplifier of the required gain.

5.12 Linear read-out amplifier for resistive bridge circuit

Fig 5.9 shows a linear readout for a resistive bridge. This type of circuit is used for matching resistor values in circuits similar to the *Wheatstone bridge*.

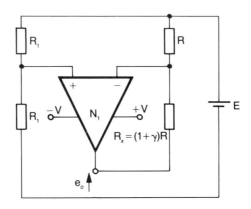

Fig 5-9 Linear readout amplifier for a resistive bridge circuit

a Assume that the resistor R_x is being checked to see whether it matches the resistor R.
b Let its value be $(1 + \gamma)R$. The voltage at the positive and negative inputs to the operational amplifier must effectively be equal, because of the negative feedback supplied by R_x.

Hence: $\dfrac{E}{2} = \dfrac{E(1 + \gamma)R + e_o R}{(1 + \gamma)R + R} = \dfrac{E(1 + \gamma) + e_o}{2 + \gamma}$.

Thus: $(2 + \gamma)E = 2E(1 + \gamma) + 2e_o$
ie $2E + \gamma E - 2E - 2E\gamma = 2e_o$

so, $e_o = E\left(-\dfrac{\gamma}{2}\right)$.

Hence, e_o is proportional to the fractional difference in resistance between the resistor under test and the standard.

c Thus, if $E = 10$ V and $R_x = 110$ Ω, while $R = 100$ Ω.

Then, $\gamma = 0·1$ and $e_o = -10$ V $\times \dfrac{0·1}{2} = -0·5$ V.

When $R_x = R = 100$ Ω, then $e_o = 0$ V. A table could be developed as follows:

Selected applications of operational amplifiers

$R_x(\Omega)$	$R(\Omega)$	γ	e_o
50	100	−0·5	2·5
80	100	−0·2	1·0
90	100	−0·1	0·5
100	100	0	0
110	100	+0·1	−0·5
150	100	+0·5	−2·5

e_o has units of −50 mV/Ω error.

5.13 Miscellaneous circuit functions of operational amplifiers

The following circuit diagrams represent a few miscellaneous functions of operational amplifiers.

- **Bass and treble tone control circuit**

The circuit shown in Fig 5.10 illustrates a bass and treble tone control circuit. The operational amplifier can be any type suitable for audio work.

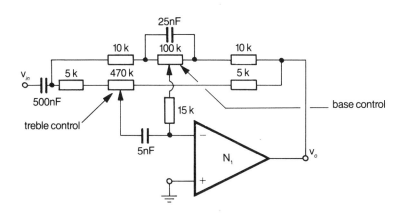

Fig 5·10 Bass and treble tone control circuit

Operational Amplifier Circuits

- **Scratch filter circuit**

The circuit shown in Fig 5.11 illustrates a scratch filter circuit. The input must have a DC path to ground. R_1 can vary from 10 k to 20 k and C_1 should be about 1·5 nF. f_C will range from 5 kHz to 10 kHz.

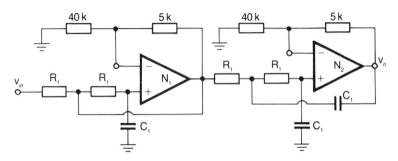

Fig 5·11 Scratch filter circuit

- **Switching power amplifier**

Fig 5.12 represents a switching power amplifier. Two 311 comparator integrated circuits may be used as the driver stage for class B amplifier. The output transistor (2N 3763) only conducts when alternate input cycles exceed a small preset threshold voltage.

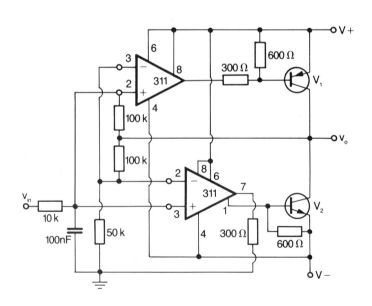

Fig 5·12 Switching power amplifier

Selected applications of operational amplifiers

- **Log converter circuit**

Fig 5.13 represents a log converter. v_o changes by 1 volt for every octave change in i_{in} current. The matched transistors can be BC 212L type.

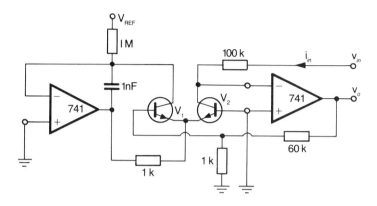

Fig 5-13 Log converter circuit

- **Musical chiming circuit**

Fig 5.14 shows a musical chiming circuit. It could be used to add a chime output to a percussion synthesiser. V_o 'rings' in response to a narrow pulse input. The amplitude of the damped oscillation peaks rapidly then decreases exponentially. Different tuning arrangements will produce varied output sounds.

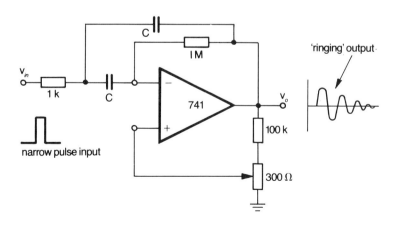

Fig 5-14 Musical chiming circuit

Operational Amplifier Circuits

- **R-S flip flop**

Many logic systems operate in a series of steps. As memories, they hold data from one period in the sequence to another. The circuits for counters and frequency dividers are similar and often one circuit can be used for both purposes. An R–S flip flop (Reset–Set) circuit is shown in Fig 5.15. Any operational amplifier may be used. $R_2 = 24\, R_1$ and $R_2 < \dfrac{V}{0\cdot 05}$.

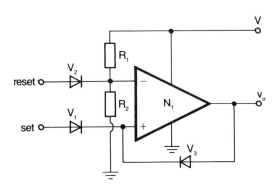

Fig 5·15 Operational amplifier R-S flip flop

5.14 Additional worked examples

Example 2

In the circuit diagram shown in Fig 5.16 determine the following:
a The rms current in R_1.
b The rms current in R_L and its phase, relative to v_s.
c The input resistance, designated as r_{in} on the circuit diagram.
d The amplifier output current, i_o, magnitude and direction.

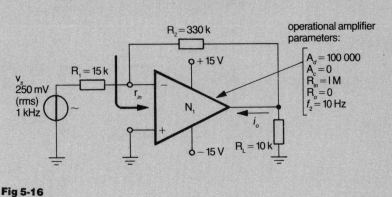

Fig 5·16

Solution

a The current in $R_1 = \dfrac{v_s}{R_1}$, since inverting input may be treated as a virtual earth point because the non-inverting terminal is grounded.

$$= \frac{0\cdot 25}{15\text{ k}} \text{ A (rms)} = 16\cdot 7 \text{ μA (rms)}$$

b $v_o = -\left(\dfrac{R_2}{R_1}\right)v_s = -\left(\dfrac{330\text{ k}}{15\text{ k}}\right)0\cdot 25 = -5\cdot 5$ V (rms)

Current in $R_L = i_L = \dfrac{v_o}{R_L} = -\dfrac{5\cdot 5}{10\text{ k}}$ A (rms)

That is, 550 μA (rms) *antiphase* to v_s.

c By *Miller's Theorem* (see a suitable text if you don't know this theorem):

$$r_{in} = \frac{R_2}{1-A_V} \| R_{in}.$$

(The effective answer is obtainable without use of Miller's Theorem.)

$$A_V = -A_d$$

Hence: $r_{in} = \dfrac{R_2}{1+A_d} \| R_{in} \simeq \dfrac{R_2}{A_d} \| R_{in}$

so $r_{in} = \dfrac{330\text{ k}}{100\,000} \| 1\text{ M} = 3\cdot 3 \text{ Ω} \| 1\text{ M}.$

Hence: $r_{in} = 3\cdot 3$ Ω approximately

ie a low impedance to ground consistent with inverting input being a virtual earth point.

d Since current, i_2, in R_2 is almost exactly equal to that in R_1, it will be 16·7 μA.

Hence, as can be seen in Fig 5.17 the amplifier output current $= i_2 - i_L$

$i_o = 16\cdot 7$ μA $- (-550$ μA$)$

so $i_o = 567$ μA (rms).

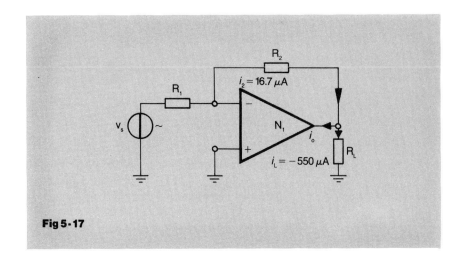

Fig 5·17

Example 3

From Fig 5.18 determine the total voltage at points A, B and C, making reasonable practical approximations. Assume C_1 and C_2 have negligible reactance at 2 kHz.

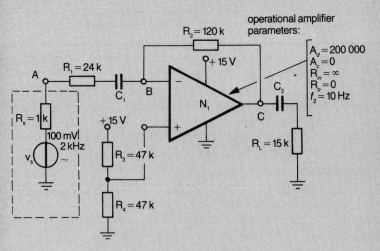

Fig 5·18

Solution

Point A
$V^+ = +7.5$ V due to the voltage divider R_3, R_4.
　　$=$ DC voltage present at $(-)$ input terminal.

The AC signal at (−) input terminal will be negligibly small by comparison with v_S.

Hence, total voltage at point A will be given by:
a The DC component is zero.
b The AC component, where the (−) input terminal may be treated as AC ground potential).

So AC component $= \left(\dfrac{R_1}{R_1 + R_s}\right) v_s = \left(\dfrac{24\ k}{25\ k}\right) 100\ mV = 96\ mV\ (rms)$

Point B
Total voltage at Point B will be given by:
a The AC component $= \dfrac{v_o}{-A_d} = \dfrac{-0.48}{-200\,000}$
   $= 2.4\ \mu V$ relative to (+ input) terminal (which is negligibly small compared with v_s).
b The DC component will be +7.5 V.

Point C
Total voltage at Point C will be given by:
a The AC output voltage, $v_o = -\left(\dfrac{R_2}{R_1 + R_s}\right) v_s = -\left(\dfrac{120\ k}{25\ k}\right) 0.1$
   $= -0.48\ V\ (rms)$
b The DC component = +7.5 volts.

Example 4

From Fig 5.19 determine the exact value of v_o, allowing for effect of R_{in} and A_d being finite quantities.

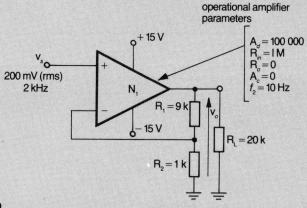

Fig 5·19

Solution

An appropriate equivalent circuit would be:

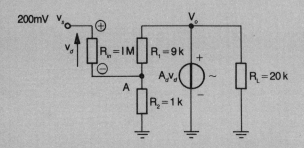

Fig 5-20

Refer to Fig 5.20.

Signal voltage at point A $= (v_s - v_d)$
so, using nodal analysis:

$$\frac{v_s}{R_{in}} + \frac{A_d v_d}{R_1} + \frac{o}{R_2} = (v_s - v_d)\left[\frac{1}{R_1} + \frac{1}{R_2} + \frac{1}{R_{in}}\right]$$

$$= \frac{0 \cdot 2}{10^6} + \frac{100\,000\,v_d}{9 \times 10^3} = (0 \cdot 2 - v_d)\left[\frac{1}{9 \times 10^3} + \frac{1}{10^3} + \frac{1}{10^6}\right].$$

That is, $11 \cdot 11222 \, v_d = 2 \cdot 22222 \times 10^{-4}$

Hence: $v_d = 1 \cdot 9998 \times 10^{-5}$ V
$v_o = A_d v_d$
$= 100\,000 \times 1 \cdot 9998 \times 10^{-5}$

Hence: $v_o = 1 \cdot 9998$ V (rms)

In other words, very close to the expected value of 2·0 V.

Example 5

The series pass transistor of a voltage regulator has the following thermal characteristics:

T_J (max) = 200°C
P_D (max) at $T_C = 25°C = 115$ W.

The regulator is intended to operate in a maximum ambient temperature of 60°C. Its unregulated supply voltage can vary from 24 to 32 volts, whilst its output is adjustable over the range 10 to 18 volts. The load resistance may be any value from 8 Ω to infinity.

Selected applications of operational amplifiers

The transistor requires a heat sink, which is to be attached to it via an electrical insulating washer with thermal resistance of 1·5°C/W. Determine the maximum permissible thermal resistance of the heat sink to ambient.

Solution

Considering the transistor, it is evident that with 115 W dissipation at a case temperature of 25°C, the junction temperature must just reach 200°C.

Hence the thermal diagram is as shown in Fig 5.21.
So $T_c + P\theta_{j-c} = T_j$ (max)
ie $25 + 115\, \theta_{j-c} = 200$
hence $\theta_{j-c} = 1\cdot5217°C/W$

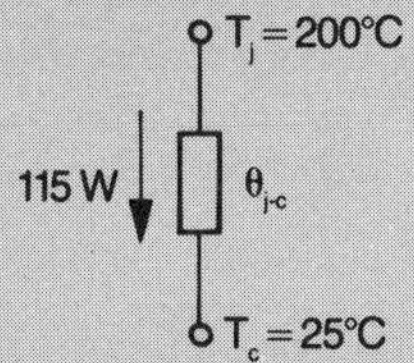

Fig 5-21

The thermal diagram for the transistor-heat sink combination will be as shown in Fig 5.22.

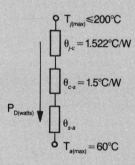

Fig 5-22

Now the dissipation in the transistor is determined by $V_{CE}I_C$ and will be greatest for a supply voltage of 32 V and a load resistance of 8 Ω.

So $P_D(\text{max}) = (32 - V_L)\dfrac{V_L}{8}$ W.

It should be evident from this relationship that if V_L is reduced, V_{CE} increases, but I_C decreases. Hence, it is necessary to determine the value of V_L which maximises P_D. There are two ways of doing this: by calculus, and graphically.

a The calculus approach involves determining when $\dfrac{dP_D}{dV_L} = 0$,

ie when $\left[(32 - V_L) \times \dfrac{1}{8}\right] + \left[(-1)\dfrac{V_L}{8}\right] = 0$.

So $V_L = 16$ V, which corresponds to a P_D of 32 W.

b The graphical approach consists of plotting P_D against V_L and noting when P_D has its maximum value.

A table showing P_D values corresponding to various V_L values is shown.

V_L(V)	10	11	12	13	14	15	16	17	18
P_D(W)	27·5	28·875	30	30·875	31·5	31·875	32	31·875	31·5

It is obvious from these results that, if they were plotted, the peak value of P_D occurs when V_L is 16 V. P_D is then 32 W.

Considering the thermal diagram shown in Fig 5.22
$T_a(\text{max}) + P_D(\text{max})[\theta_{s-a} + \theta_{c-s} + \theta_{j-c}] \leq T_j(\text{max})$
so $60 + 32[\theta_{s-a} + 1·5 + 1·522] \leq 200$;
ie $\theta_{s-a} \leq 1·353°\text{C/W}$ or $\theta_{s-a}(\text{max}) = 1·353°\text{C/W}$.

Example 6

An electrically ideal complementary symmetry amplifier has its two matched output transistors fixed to a common heat sink via electrical insulating washers with thermal resistance of 1·4°C/W. The transistor's maximum junction temperature is 200°C and their thermal resistance, junction to case, is 2·0°C/W. The circuit operates from a 75 volt supply rail and delivers its output power to a 10 Ω load.

The heat sink has a thermal resistance of 3·5°C/W and the amplifier is intended to operate in ambient temperatures up to 50°C.

a Determine the maximum power output which the amplifier can deliver to the 10 Ω load under maximum ambient temperature conditions.

b During manufacture, one amplifier has one of its output transistors incorrectly assembled to the heat sink, so that the thermal resistance from case to sink becomes 4°C/W instead of 1·4°C/W. Determine the power output level at which the amplifier should theoretically suffer thermal failure.

Solution

a $P_o(\text{max}) = \dfrac{V^2_{(\text{rms})}}{R_L} = \dfrac{\left[\left(\dfrac{75}{2}\right)\dfrac{1}{\sqrt{2}}\right]^2}{10} = 70 \cdot 31 \text{ W}$

b The thermal diagram of the faulty amplifier is shown in Fig 5.23.

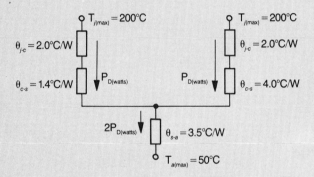

Fig 5-23

The incorrectly assembled transistor will suffer thermal failure first, when
$T_a(\text{max}) + (2P_D \times \theta_{s-a}) + P_D(\theta_{c-s} + \theta_{j-c}) > 200°C$
ie $50 + (2P_D \times 3 \cdot 5) + P_D(4 \cdot 0 + 2 \cdot 0) > 200$
$P_D > 11 \cdot 54$ W
Now $P_{DC(av)} = P_o + 2P_D$, ie $V_{CC} \times I_{DC(av)} = I^2_{\text{rms}} R_L + 2P_D$
so $V_{CC} \left(\dfrac{I_{(\text{rms})} \sqrt{2}}{\pi} \right) = (I^2_{\text{rms}} R_L) + 2P_D$
so $I^2_{\text{rms}} R_L - \dfrac{V_{CC} \sqrt{2} \, I_{\text{rms}}}{\pi} + 2P_D = 0$
$10 \, I^2_{\text{rms}} - 33 \cdot 76 \, I_{\text{rms}} + 23 \cdot 07 = 0$
Solving this quadratic equation for $I_{(\text{rms})}$ gives values.
ie $I_{\text{rms}} = 2 \cdot 42$ or $0 \cdot 952$ A.
It then follows that P_o must be $I^2_{\text{rms}} R_L$,
ie $58 \cdot 8$ W or $9 \cdot 06$ W.

The transistor failure will occur at 9·06 watts output because this output level will be reached before 58·8 watts. However, why are there two answers? Do both in fact give the same transistor dissipation?

At $P_o = 9\cdot06$ W $I_{rms} = 0\cdot952$ A.

The value of P_{DC} will be $V_{CC} I_{DC(av)}$, ie $V_{CC} I_{rms}\left(\dfrac{\sqrt{2}}{\pi}\right) = 32\cdot1$ W

so $P_D = \dfrac{P_{DC} - P_o}{2} = \dfrac{32\cdot1 - 9\cdot06}{2} = 11\cdot5$ W.

Similarly it can be shown that when $P_o = 58\cdot7$ W. $P_D = 11\cdot5$ W. Hence, there is exactly the same dissipation in the transistors. It should also be noted that both values are otherwise acceptable because they are both power outputs which are less than $P_o(max)$ ie $< 70\cdot3$ W.

INDEX

A_{CL}, 6, 23, 40, 41
adder circuit, 12
adjustable gain, 117, 118
amplifiers
  bridge, 87, 92, 93
  classification, 76–9
  complementary symmetry, 84, 85
  differential, 25
  gain, 25, 117, 118
  integrator, 24
  power, 76, 85, 86, 90, 91, 92
  push–pull, 80–4
  unity gain, 16, 25, 26
  voltage follower, 16, 25, 26
A_{OL}, 11, 35, 40, 41
astable multivibrator, 51
audio circuits, 111
A_{VO}, 1

bandwidth, 1, 33, 34, 35, 39, 40
base and treble tone control, 121
biasing arrangements, 106–08
bistable multivibrator, 51–3
bridge amplifier, 87

CMRR, 3, 4
common mode voltage gain, 3
crossover distortion, 82, 83
current limiting circuit, 102
cut off frequency, 34

DC voltmeter, 112, 113
differential amplifier, 12
differential amplifier gain, 3
differential gain, 11
differential input signal, 3
differentiator, 15, 16
drift, 14
dual-in-line package, 20, 21

EMG monitor, 113, 114
error components
  bias currents, 14
  offset current, 14
  offset voltage, 14
  slew rate, 42–7

feedback, 5, 21, 11
filter applications, 119
flip-flop multivibrator, 51
free-running multivibrator, 53, 54
frequency response, 33

gain
  adjustable, 117, 118
  common-mode voltage, 3
  differential, 3
  unity, 16, 25, 26
gain bandwidth product, 34
generators, *see* signal generators

heat sink, 93–5
high frequency cut off, 39–40

IC regulators, 96–104
ideal op amp characteristics, 1, 6
input bias current, 14
input offset current, 14
input offset voltage, 14
input resistance, 1
input signal variation, 35
integrator, 14, 15
internal frequency compensation, 33
inverting operational amplifier, 5–7

light intensity circuit, 118, 119
load lines, 77–9
log converter circuit, 123
low cut off frequency, 34

monostable multivibrator, 51
meter resistance, 112, 113
musical chiming circuit, 123

negative feedback, 5, 21, 22
nodal analysis, 28–30

offset current, 14
offset voltage, 14
one-shot multivibrator, 51
operational amplifier regulator,
  96–104

Operational Amplifier Circuits

operational amplifiers
  advantages, 1, 2
  basic terminals, 2–5, 8
  bias currents, 14
  circuit symbols, 4
  closed loop gain, 6, 23, 40, 41
  differential input, 3
  drift, 12
  frequency compensation, 33
  frequency response curves, 36–8, 48
  gain-bandwidth product, 34
  ideal, 1
  input power, 35, 36
  inverting, 5–7
  multivibrators, 53–8
  non-inverting, 7–9
  output power, 35, 36
  output resistance, 1, 40, 41
  packages, 20, 21
  parameters, 11
  power supply, 96–104
  ramp generator, 58
  sawtooth wave generator, 61, 62
  sine wave oscillator, 62–4
  slew rate, 42–7
  terminals, 20, 21
  triangular wave generator, 59–61
  unity gain bandwidth, 16, 25, 26
  voltage follower, 16, 25, 26
oscillation, 47

packages, 20, 21
peak detectors, 116, 117
phase angle, 37, 63
phase shift, 48, 62, 63
phase splitter, 84
photodiode, 114, 115
positive feedback, 47, 62–4
power amplifiers, 76, 85, 86, 90–2
power supply, 96–104
power transformer, 78–81
precision rectifier, 111, 112
pre-regulation, 100, 101
programmable unijunction transistor, 62
push–pull amplifier, 81–4

ramp generator, 58
rectifiers, 111, 112, 115

regulated power supply, 96–104
resistive bridge, 120, 121
rise time, 38, 39, 48, 49
roll off, 34
R–S flip flop, 124

sawtooth wave generator, 61, 62
scratch filter circuit, 122
signal generators
  astable, 51
  bistable, 51–3
  monostable, 51
  one-shot, 51
  ramp, 58
  sine wave, 62–4
  square wave, 51
  Wien bridge, 65, 72, 73
slew rate, 42–7
summing amplifier, 12, 13
superposition theorem, 26–8
switching power amplifier, 122

terminals, 20, 21
thermal protection, 93–5
three terminal regulators, 103
triangular wave generator, 59–61

unity gain bandwidth, 16, 25, 26
unregulated power supply, 96–104
unstable state, 53–8
upper cut off frequency, 39, 40

variable gain, 117, 118
virtual ground, 12
voltage follower, 16, 25, 26

waveform generators, *see* signal generators
Wien bridge oscillator, 65, 72, 73

zener diode, 60, 61, 74, 96–101